THE
INTELLIGENT
INVESTOR:
SILICON
VALLEY

"For everyone who has arrived at a place where they have the resources to back capable technology entrepreneurs who are changing the world for the better"

Works by Alison Davis and Matthew C. Le Merle

The Fifth Era

Build Your Fortune in the Fifth Era

Corporate Innovation in the Fifth Era

Blockchain Competitive Advantage

The Intelligent Investor: Silicon Valley

Works by Matthew Le Merle

Second Chance: A novel

The Ministry of Bitcoin: A novel

Woodland
Vacaville
Napa
Fairfield
Concord
Petaluma
Vallejo
Novato
Berkeley
San Francisco
Baia di San Francisco
San Mateu del Maestrat
Palo Alto
Sunnyvale

THE INTELLIGENT INVESTOR: SILICON VALLEY

Practical wisdom for investors and entrepreneurs from 50 leading Silicon Valley angels and venture capitalists

Alison Davis
Matthew C. Le Merle

Fifth Era Media
4545 Paradise Drive,
Tiburon, CA 94920, USA

www.FifthEraMedia.com

415-994-4320

Originally published in the USA in 2020

By Fifth Era Media.

Fifth Era Media is a registered trademark of Fifth Era, LLC.

Library of Congress Control Number: 2020946593

ISBN: 978-1-9502-4812-4 (paperback)

ISBN: 978-1-9502-4813-1 (hardcover)

ISBN: 978-1-9502-4811-7 (Ebook-EPUB)

Compilation: Leonardo Q. Le Merle

Cover and interior design: Andy Meaden meadencreative.com

Contents

FIFTH
ERA

Preface – The Fifth Era

After almost 200 years living in the Industrial Era, humankind is now entering a new era driven by the twin forces of the digital and life sciences revolutions. Today we are all connected, and nearly all the world's information is online. Every industry has been impacted, and many have gone through fundamental transformations. We have also begun not only to edit and adjust our plants and animals, but are beginning societal discussion as to whether to alter the very nature of human beings.

It is a time of great opportunity, and also great uncertainty and even fear. It is a time of unprecedented disruption and creative destruction, and also a time of enormous value and wealth destruction and creation - the greatest the world has ever seen. We call this the Fifth Era, and in our previous books—*The Fifth Era; Build your Fortune in the Fifth Era; Corporate Innovation in the Fifth Era; and Blockchain Competitive Advantage*—we have written about how individuals and corporations can prepare to thrive rather than be left behind in this new era.

In these books we detail the essential changes underway. We make the point that this is not just a matter of one or two major technological shifts. Rather, we are living in a time when within a few decades an unprecedented number of important and compounding innovations—the Internet, artificial intelligence, the Internet of things and the sensor revolution, 3D manufacturing and the distributed maker movement, augmented reality, clean energy technologies, gene editing, quantum computing, blockchain, and many others—are all coming of age in the same timeframe, propelling us into a completely new era of human life on planet Earth.

It is still early in the development of these new technologies and the business innovations that they can and will drive. Much will change, every industry will be impacted, and in the process many projects and players will fail. Entrepreneurs will work hard only to find they can't get traction before their resources run out. Investors will lose money more often than they see a return even if the overall return is an attractive one. Established companies will have varying degrees of success in adapting and retaining their positions. Trillions of dollars of value will be lost and gained – even more so than 20 years ago when we first rolled out the Internet and connected the world. Several of the world's largest companies today by market capitalization didn't exist 25 years ago, and others that were much admired are not relevant any longer.

We believe we are transitioning from the Industrial Era into a new Fifth Era in which a global digital world will transform everything humans do. Those companies that leverage new technologies to create new business models and customer offerings and build strong innovation muscle and capabilities will create enormous market value and those that don't will flounder. We are living in unprecedented times in which technology and innovation are the leitmotifs of our lives and power our global economy and the human experience. This is the core thesis of our previous books. If you are interested in exploring further, we invite you to download a **free** copy of our introductory book *The Fifth Era* which can be found at **www.fifthera.com**

Introduction

In 2010 we noticed that Apple had joined Microsoft as one of two west coast-based technology companies on the list of the world's most valuable companies – a significant change since for the prior few years only Microsoft had been able to hold its place on the list. The rankings were still dominated by the legacy of the 'Industrial Era' all too clearly, but technology-based companies were starting to climb up the rankings again.

2010 World's Largest Companies by Market Cap

PetroChina ($329 bn)

Exxon ($316 bn)

Microsoft ($256 bn)

ICBC ($246 bn)

Apple ($213 bn)

BHP Billiton ($209 bn)

Wal-Mart ($209 bn)

Berkshire Hathaway ($200 bn)

General Electric ($194 bn)

China Mobile ($193 bn)

We both had early careers as partners at top management consulting firms serving established large companies in legacy financial services, consumer products, technology and other industries. When we moved from New York to the Bay Area just as the world wide web was becoming a commercial reality, we increasingly began to serve new technology clients like Amazon, Cisco, eBay, Google, PayPal and many others. At the same time we both became active early-stage technology investors and began to get good exposure to the explosion of technology startups in the Silicon Valley ecosystem. It was in 2010, looking at the valuation rankings and seeing the incredible growth trajectory of many new technology companies, we started to believe the tailwinds behind new technologies were irreversible and that – over the next few decades – more and more of the most valuable companies in the world would be those leaders in innovation leveraging these emerging technologies.

By then we were writing and speaking more about the importance of new technologies and the innovation economy. We would often ask people to guess which companies were in the top 10 of the world's most valuable companies. Even by 2015, very few people got it right, mostly assuming the energy, banking, and industrial companies of the past were still dominating the rankings. When we put up the actual 2015 data, it was a surprise to most. The three most valuable companies in the world were all west coast based technology companies!

2015 World's Largest Companies by Market Cap

Apple ($621 bn)

Google ($408 bn)

Microsoft ($347 bn)

Berkshire Hathaway ($318 bn)

Exxon Mobile ($304 bn)

Johnson & Johnson ($257 bn)

General Electric ($248 bn)

China Mobile ($243 bn)

Novartis ($240 bn)

Nestle ($233 bn)

We wondered what Benjamin Graham – the great author of the world's top selling book on investing *The Intelligent Investor* – would have made of this? Graham's book was written in 1949 and still outsells other books as an authority on the science and discipline of investing. In 1949, public technology companies were mostly poor performers, and Graham thought so little of the sector that he relegated his discussion of it to one page in a final Appendix 7 where he concludes *"The phenomenal success of IBM and a few other companies is bound to produce a spate of public offerings of new issues in the fields, for which large losses are virtually guaranteed."* That's it! In 600 pages, that is all he had to say about investing in technology companies.

No wonder then that the great proponents of Benjamin Graham and value investing – Warren Buffet and Charlie Munger – have singularly and repeatedly emphasized their lack of support for technology investing. Even though they have excelled at investing in traditional companies, and Berkshire Hathaway has maintained an impressive ranking among the world's most valuable companies, technology investing is just not their thing. Ironically, Buffet broke with this long held view in 2016 to buy into Apple and today Apple has powered an incredible portion of the performance of an otherwise sclerotic Berkshire Hathaway (For the last decade, their value investing strategy has underperformed the S&P 500 even with the inclusion of Apple).

However, returning to Benjamin Graham, he was first and foremost an empiricist. So what would he make of this data from July 2020? :

2020 World's Largest Companies by Market Cap

Apple ($1,576 bn)

Microsoft ($1,551 bn)

Amazon.com ($1,432 bn)

Alphabet ($980 bn)

Facebook ($676 bn)

Tencent ($620 bn)

Alibaba ($579 bn)

Berkshire Hathaway ($433 bn)

Visa ($413 bn)

Johnson & Johnson ($370 bn)

We think *The Intelligent Investor 2020* would be a very different book. If Benjamin Graham was doing his work today, instead of segmenting industry verticals into 'value' versus 'laggard' stocks, he would have no option but to see that most of the market value of the last decade has been created by companies built on innovation and new technologies. Benjamin Graham would have won by just dividing the public market into two parts and investing in the one that contained new technology based companies. We believe traditional forms of corporate valuation based on long term discounted cash flow models can lead to blind spots and myopia in the current context. For example, most of the value in DCF models is in the terminal value, and when industries and markets are changing and being disrupted so quickly, we question the value of assuming a stable growth rate in perpetuity for any company in any industry. The venture capital industry and early stage technology investors are far closer to understanding how to value and invest in the high growth technology and innovation sector than traditional finance theory.

We believe it is time for a new book on

intelligent investing. It seemed to us though that rather than emulate *The Intelligent Investor* and try to develop a new quantitative methodology to apply to disruptive technology investing, it would be more powerful to 'crowd source' insights and collect wisdom from those at the center of the Silicon Valley innovation ecosystem, those at the forefront of committing their own financial and human capital to intelligent investing in the new economy.

So we have reached out to 50 experienced, successful and insightful Silicon Valley angel investors, venture capitalists, founders, incubator and accelerator operators, and technology service providers – a broad and diverse set of inspiring people with different lenses – and we asked them for their views in a series of questions. Their responses have created this book – an empirical study where we capture their wisdom, letting their voices do the talking, rather than trying to edit or synthesize their words. We found it truly insightful as well as inspiring. We hope you do too.

Alison Davis
Matthew C. Le Merle
San Francisco, California, USA

"Every Failure gets you a few steps closer to your Success."

Michele Ellie Ahi

Angel Investor, Venture Capital Partner, Entrepreneur's Mentor

Keiretsu Forum, AIV Ventures, SCU Ciocca Center for Innovation and Entrepreneurship

What did you want to be as a child?
Doctor or a Data Scientist.

What was your first career?
Software Engineer.

What other careers have you had before your current one?
I became a Computer Science Engineer and started working as a software engineer – Now I teach, mentor, and invest in technologies but I am still very hands-on.

Why do you choose to focus on early stage technology investing?
I like to be there when it all starts! I like the challenge, and of course, like the low valuations of early stage. I normally invest on the team and the technology.

When your investing is not going well, what do you do or say to yourself to keep motivated?
My investments are very diversified. I normally set my expectation low; when my investment is not going well – I first try to suggest a few ideas and offer to help (that is why a coachable solid team matters.) But at the end we all know any investment is risky.

What do you love most about what you do now?
The learning…I learn when I invest; I learn when I teach; and I learn when I mentor.

What dreams do you have for the next 10 years?
I have started a new company, which builds products and tools for the healthcare industry – I would like to expand that. Healthcare more than ever needs all of the innovation possible. The first step is patient education. I would like to continue my investments and teaching. I am getting my kids involved in AIV Ventures; encouraging them to explore, invest, and execute their ideas with a right team!

If you could change one thing about the world we live in, what would it be?
Transparency. The World would be a much better place if every country, government, citizen is transparent, by default. The First Act of transparency is providing "the RIGHT information."

What was your first technology investment and what happened?
I invested in a technology, which was about Personalized AI-Powered Search Engine on the go – it was back in 2004 – and the technology was too early for the 2004 market. Back then the only semi smart mobile phone was the Blackberry. So IT is important to be futuristic – but also to be realistic and know the markets very well for which you are trying to build products. That is why we need both R&D; **R**esearch for being forward thinking and **D**evelopment for building products on time for the current market.

Can you share a story of one of your best investments, and why it went so well?
I invested in a company (investment and also got involved as a technology advisor). We built an IP, which was about creating smart job boards and candidate matches; we added smart algorithms to profile the candidates, perform a creative gap analysis to find the best matched candidate - per the skill set of the job (soft skills and hard skills). The technology was acquired by another corporation. It is doing well and currently providing services to companies such as Twitter and Uber.

Can you share a story of one of your worst investments and why it went so badly?
I invested in a company per one of my friends' referral – The company was based in the EU. The CEO/Founder was planning to move the product to the USA. The product was a Mobile Shopping App. It was a fun app and had a simple gamification feature that made shopping more fun, but it needed to be

localized for the US market. Even though the CEO seemed coachable at first. He resisted the very first important suggestions made by the US advisors/investors and eventually he decided to delay the launch in the US. The company and product are still running in EU but never made it to the US! It was a good lesson for me, and a reminder: Invest in the Team (Founder/CEO) and then the technology/IP, then the rest (product, valuation, biz model, rev, market etc.)

In your opinion, what is so special about Silicon Valley?

You cannot find the entrepreneurship energy that we have in Silicon Valley anywhere in the world. This entrepreneurship energy/go-get-it mentality could be due to:

- The competitive environment
- The cutthroat environment
- The high cost of living
- The expensive housing market
- The egos of people who live here
- San Francisco and the tourists who visit us, and dream of making it here
- The great education and colleges in Silicon Valley, such as (Stanford, USC, UCLA, Berkeley, etc.)
- It could be that the big names all started here such as HP, Intel, Apple, Google, and so on.

Who have been role models you most admire?

Bill Gates and his wife – they are very accomplished and can do so much with their $ billions, but instead through their Foundation, they have been spending so much on helping people and caring about the world. They truly walk the talk!

What is the best advice you ever received about how to succeed in Silicon Valley?

About 18 years ago from my CEO when I was a senior director and was worried about if we can sell our products, back then during the 2001/2002 crisis. He said *"Ellie just let us work on the product and build it well but do not worry about the market; we should not follow the money. If our product is good, money will follow us"*

What do you believe are the most important traits of a technology entrepreneur?

The most important trait of an entrepreneur is to surround herself/himself with a great team – a great team that is smarter than him/her and is as passionate about the technology as she/he is.

What do you believe are the most important traits of an early stage technology investor?

Understand the technology and the market the team is targeting, and most importantly believe in the team – the team must have a good track record or, if it is a young team, they must be passionate about what they are building and be in it for all the ups and downs. I am an early stage investor and am very careful not to invest in those who are in it for quick fame or money/exit. I like to see that the early stage startup is thinking in terms of building a company rather than creating a company/brand and a quick exit!

If you could tell a technology entrepreneur just one thing, what would it be?

Less is more. Start small – focus on your core technology and the main problem you are solving; execute it RIGHT, design it to be SCALABLE - with a solid plan to expand both horizontally and vertically.

What do you know now that you wish you had known earlier on?

Consult more to/with my investor colleagues. Study the market better – Best product will not make it if the market is not ready for it.

Would you please create a personal quote that captures important wisdom about participating in early stage investing?

"Every Failure gets you a few steps closer to your Success."

I love and personally follow this quote. 17th century Japanese poet and samurai, Mizuta Masahide:

"My barn having burned down, I can now see the moon."

The best investors have a unique combination of instinct, deep domain expertise, and the ability to make people feel safe and supported. The challenge is those are rare qualities in and of themselves, let alone in combination. Now add in the need to have a very high-risk tolerance and extraordinary physical and emotional endurance, and you begin to see why there are few truly great early stage investors.

"Be sure to take a portfolio approach... Remember there is value even in those that fail."

Bodil "Bo" Arlander

Angel Investor, Venture Capital Partner

Golden Seeds, Portfolia Active Aging & Longevity Fund

What did you want to be as a child?
Medical researcher.

What was your first career?
Fashion model.

What other careers have you had before your current one?
Investment Banker, Private Equity Partner.

Why do you choose to focus on early stage technology investing?
Helping entrepreneurs with both funding and advice is very rewarding. Being able to see emerging brands and technologies is a bonus.

When your investing is not going well, what do you do or say to yourself to keep motivated?
I only invest dollars I can afford to lose and take a very long-term approach, so keeping my optimism isn't all that hard.

What do you love most about what you do now?
It offers me freedom to do what I want when I want while still feeling like I can contribute and make a difference in the world.

What dreams do you have for the next 10 years?
I hope that technologies that make the world a cleaner, healthier and more sustainable place for all get the funding and support they need to succeed. I also hope that global politics don't get in the way of doing what's right for the planet so that future generations can enjoy this beautiful place we live in.

If you could change one thing about the world we live in, what would it be?
I wish there was more focus, efforts and dollars dedicated to making sure our planet remains inhabitable and a safe place for generations to come. This also means that those countries that have the know how and resources to lead these efforts need to share the knowledge and help guide those who don't without allowing politics to get in the way. We all share this one planet and it's the only one we've got so we should all care about it and take care of it to the best of our abilities.

What was your first technology investment and what happened?
The first technology investment I was exposed to was back in 1999 during my tenure at a private equity firm based in New York. We had been alerted to a potential going private opportunity in the integrated circuit space and ended up pursuing it in partnership with another private equity firm that had much deeper experience in technology investing than we did. The company had suffered a severe drop in its share price as a result of an effort to branch into a new space that turned out to be very competitive and a much smaller opportunity than the company had expected. The investment thesis was that by refocusing the company's business on its core products, it could get back to successfully growing again. In the transaction, we backed the existing COO, who was extremely talented, and his terrific management team who were able to grow the company's already dominant market share in its core product through a relentless focus on innovation. We were in the investment for only two years but ended up with a tremendous return on investment when we sold our stake a year after it was taken public again.

There is a lot more focus now by investors, employees and customers on companies that do good (and do no harm) as well as make money. How do you think about this topic in the context of your investing?
As I consider my own career as an investor, first within private equity and more recently as an angel and VC investor, there has clearly been a shift in my awareness of and focus on factors other than pure dollar investment returns. I attribute this largely to the frequent and wide dissemination of news related to pollution, global warming and its consequences, exploitation of labor forces and unsafe

working conditions in many countries as well as discrimination against individuals based on sex, race or religion.

While as an investor I obviously strive to make profitable investments, these days I am keenly interested in how good a corporate citizen the entity is that I am considering investing in. From a pure investment perspective in regard to consumer companies, I think today's consumer, especially the younger generations, are much more discriminating in their purchase behavior in that they expect the companies they buy from to use sustainable materials, employ ethical manufacturing practices, to waste less resources, to employ a diverse workforce and support causes that help the under privileged or support the planet. This means that those consumer companies, new or old, that stick to past ways of doing business with little regard to the footprint left behind are unlikely to succeed and are therefore unlikely to be good investments.

In addition to using this as an important filter when considering new consumer investments, I have expanded my investment universe to include other industries in order to be able to have a broader impact personally. Whereas in the past I was heavily focused on just consumer facing companies, I am nowadays on the lookout for cool new companies regardless of industry that approach problems in the way we currently operate and attempt to offer better mousetraps that do less harm to the planet.

What do you believe are the most important traits of a technology entrepreneur?

There is a Finnish word "sisu" which encompasses the characteristics of most truly successful people I know regardless of their field. It is the ability to forge ahead in the face of adversity and near impossible odds – a "never say die" attitude. Without such grit and determination, which also imply a serious commitment to the cause whatever it might be, I believe it is difficult to succeed as an entrepreneur in any field.

What do you believe are the most important traits of an early stage technology investor?

I believe early stage tech investors need a lot of patience, an open mindset and an innate curiosity as to what is happening in the world around them.

If you could tell a technology entrepreneur just one thing, what would it be?

Start fund raising your next round at least six months earlier than you think you need to because chances are you'll run out of funds sooner than you expected and it will take you longer to raise the funds than you planned on.

What do you know now that you wish you had known earlier on?

There's a reason consumer products often struggle with early stage fundraising. They face a longer road to profitability and scale than most technology companies, which when you add the failure rate of start-ups in general, simply make them less attractive from an expected returns perspective.

Would you please create a personal quote that captures important wisdom about participating in early stage investing?

"Be sure to take a portfolio approach, be patient and don't expect all of your investments to turn out well. Remember there is value even in those that fail."

As Yoda says: *"Failure the greatest teacher is."*

Early stage technology investing provides a sneak peak of what the future could and probably will look like. Every day is intellectually stimulating.

"You just gotta learn to roll with the punches."

Andrew Blum

Strategist, Coach, Consultant, and Advisor

Founder & CEO of The Trium Group

What did you want to be as a child?
Happy – and the first career I ever really wanted was to be a Marine.

What was your first career?
My first career was initially as a Marine and then a Marine Officer.

What other careers have you had before your current one?
Before founding Trium, I was a traditional Management Consultant and also spent some time in software sales, as well as being an Officer in the Marine Corps.

Why do you choose to focus on early stage technology?
I am not a technology investor but as a resident of the SF Bay Area, I have been working with early stage companies for many years. They are dynamic, interesting, and always in need of the kind of support Trium provides in order to reach their growth objectives and are often willing to innovate to achieve that.

When your work is not going well, what do you do or say to yourself to keep motivated?
I always expect that the process will be difficult, and I stay in the process because it is only through confronting these difficulties that you learn and grow. When things are going well, I have learned that they will eventually not go well and vice versa. There is a saying from the Grateful Dead that goes: "*When life looks like Easy Street there is danger at your door.*" This is very true about Silicon Valley and knowing that progress isn't a sign of sure success, I also know that failure is often a natural precursor to breakthrough and success.

What do you love most about what you do now?
I am able to live a mission of making the world a better place by changing the way business leaders think. I take on deeply intellectual and emotional challenges in helping my clients that force me to evolve as a human and leader. It is a highly virtuous cycle of making a difference, learning and growing, always being challenged, and leading a thriving business with a wonderful mission-driven team.

What dreams do you have for the next 10 years?
I am nearing the later stages of my career. In the next ten years, I would love to find a path and rhythm that is more moderate. Working in Silicon Valley is extreme. By moderate, I mean balanced – where I have time to do my spiritual development, my physical work, time with family and friends, while still having a meaningful impact in the client world. All too often, it is a choice to be in the game or out of the game. Being in the game is exhausting and being out of the game is boring or rudderless so I'm hoping to find that path, the razor's edge, where I am engaged by the work but not consumed by it.

If you could change one thing about the world we live in, what would it be?
It really would be removing the levels of greed, ego and fear that are running all of the systems we are living in. The "ego mind" is out of control in Silicon Valley and "that mind" believes that it never has enough and there always has to be more.

I wish every human being could live with less fear and anxiety, and more compassion for Self and Other. Out of that consciousness, almost all problems can be solved, whether they be around climate change, social equality, or interpersonal relationships.

The underlying consciousness of human beings needs to evolve. We are still living in a very limbic response and in a world where that limbic response is largely inappropriate and the pressure of Type-A personalities fueled by venture capital and competition, and unbelievably high expectations, is at the heart of all of the innovation and all of the suffering in Silicon Valley.

What was your first technology investment and what happened?
The first interesting hot startup we worked with was the Industry Standard. At the time that was the "Economist" for the Internet economy in the 2000 boom. It was fascinating because it was a rapidly growing organization that essentially was the voice of

the new Silicon Valley at the time. John Battelle was the CEO and they engaged us to help them manage the challenge of scaling and growth while they were also, in many ways, editors of the world of scaling and growth as their judgments and commentaries about companies were read by all. Like many companies in that phase, they were on a tremendous growth curve, but it wasn't sustainable. When the dotcom bubble first burst, their advertising revenue dropped through the floor and they ultimately couldn't sustain. We learned that the voice of Silicon Valley rises and falls with the companies it reports on.

Can you share a story of one of your best investments, and why it went so well?

Since I'm not an investor per se but rather a consultant, strategist, and coach, the best investment is measured not in financial returns but in the strength of relationship built from a single client. I started working with Carl Eschenbach who was the head of the GTM organization and President of VMware. Out of that relationship spawned many different clients. I worked with his team extensively and as they eventually left VMware, they hired us directly or through referral at companies including NewRelic, Puppet, Zendesk, Cohesity, UiPath, Sequoia, Andreessen Horowitz, and others – all as an outcome of working with Carl and his team. We were able to work with Carl and his team to unlock their next level of performance. At the time, VMware was growing because they had extraordinary product market fit with their virtualization offering. As that market became mature and saturated, their team needed to evolve radically and become much more collaborative and work to sell a much broader range of products and services. Because we were able to help them do that and in a way that awoke the potential of the team and leaders within it, all of those leaders had a peak experience which they remembered and valued, and then called Trium on an ongoing business for many years after.

Can you share a story of one of your worst investments and why it went so badly?

Nearly every investment we made in the first dotcom boom, where we took equity for services – failed. It was the early days of the Internet and there was a land-grab mentality driven out of a sense of scarcity and urgency that led to many dysfunctional behaviors and ultimately business failures. The driving orientation was that there would be a "leader"

in every market space – but only one, and therefore the strategy was to grow as quickly as possible before having clarity on the product market fit or the value proposition of the organization. It was very much a gold rush mentality.

I will never forget the moment in which George Shaheen, the CEO of Anderson Consulting at the time, went on to become the CEO of Webvan. This was seen as evidence that the future was in companies that were changing the world. Ironically, it took nearly a decade for the concept of Webvan to work and you can see shadows of it in Walmart and Amazon. In those early days, the metric of success was how many people you hired and how fast you grew your top line, and that orientation has remained until recently when profitability suddenly does matter. The model at the time was grow and then figure out how we're going to make money.

In your opinion, what is so special about Silicon Valley?

The thing that makes Silicon Valley so amazing is that the right idea at the right time with the right investors can have enormous returns. When I think about it, if you are at the top of Goldman Sachs or even a Fortune 50 company, you have a chance of making 15-20 million dollars in a year. There are leaders in Silicon Valley who have an exit who make 100 million dollars or much more after a few years. This potential for outside returns creates an almost Las Vegas orientation where it's worth a chance because if you win, you win BIG. Just like in Las Vegas, people also gamble forgetting that you lose 98% of the time but they seem to only reference those people who won. This creates a lot of both interesting behavior and also dysfunctional behavior. It drives an incredible expression of energy in which people work tirelessly but that same energy is often misplaced, and people chase a dream that is unrealistic and the pursuit of which can undermine every aspect of your life.

Why do you think the innovation economy/ Silicon Valley has a poor track record on diversity and inclusion? What needs to change?

In truth, Silicon Valley is among one of the least diverse places in the world for one primary reason and that is its DNA. In the early days and still largely true today, many startup leaders have come out of the Stanford Engineering or Mathematics programs. That very narrow funnel always had a very distinct profile that is the classic geeky type guy who would go on to hire his friends and start a company. That became a self-fulfilling pattern because investors always saw that as a great profile – if I have a couple of math and engineering kids from Stanford or MIT, that's generally been a good bet. There was never really an impetus for diversity and inclusion and there is, in fact, an implicit success model. That model worked for a long time and to this day, there continue to be young Stanford kids starting companies and getting investment than any one other model. It is only through extraordinary effort and by openly challenging that original DNA, will investors begin to create the diversity and inclusion needed to really be effective in building companies that can last and thrive through many different cycles. The good news is that seems to be happening now.

There is a lot more focus now by investors, employees and customers on companies that do good (and do no harm) as well as make money. How do you think about this topic in the context of your investing?
Again, we are in the business of strategic counsel, coaching and support, and not investing per se. When we take on new clients, we take a bit of a Hippocratic oath and avoid the business of evaluating whether our clients are "worthy" of our help. Often, we see that it is our mission to work with leaders who are un-evolved as much as with leaders who are evolved. In all cases, we are trying to help them be more conscious and humane. In the past decade, nearly every company has tried to tie their work to the mission of "making the world a better place" – some do, and some don't. Ultimately, in a world of indirect and unintended consequences, it is truly hard to know whether a company supports the greater good or not. Is Uber evil because of their practices or are they moral because of the benefits they provide? What about Google? Facebook? In truth, it is hard to make this evaluation, so our work is to help leaders lead with more clarity, precision, self-awareness, and authenticity.

Who have been role models you most admire?
My role model who I most admire is Byron Katie. She is relentless in her commitment to her mission that is to create peace for everyone. Her core focus on dealing with the mind and how we perceive reality is the key factor in all change – the issues of sustainability, racial equality, balance, and all of the things that we struggle with as a society and as a

Silicon Valley community are coming out of how we think about reality. Byron Katie is focused on the core issue and lives it with absolute integrity.

What do you believe are the most important traits of a technology entrepreneur?
What is really interesting is that the key traits have gone through a massive evolution in the past year. It used to be that the key traits were brilliance combined with naked ambition and an over-expanded view of one's role in the world and their potential for impact. That did lead to a lot of great companies being founded with a very powerful and often justified egoic orientation. With the decline of WeWork and some of the stumbles we've seen at Uber and other companies, we are now looking for a different set of traits. Brilliance and bold vision are important but now need to be combined with humility, grace, humanity, and true accountability. That is the healthy evolution that has taken place and is occurring across industries.

What do you believe are the most important traits of an early stage technology investor?
The best investors have a unique combination of instinct, deep domain expertise, and the ability to make people feel safe and supported. The challenge is those are rare qualities in and of themselves, let alone in combination. Now add in the need to have a very high-risk tolerance and extraordinary physical and emotional endurance, and you begin to see why there are few truly great early stage investors.

If you could tell a technology entrepreneur just one thing, what would it be?
This is not an original statement, but it cannot be said strongly enough: Make sure you have product market fit. No matter how big your idea is, if you don't have a lot of people who are very interested in it, it won't be successful. I look at Robinhood and their history – they had enormous interest just on the idea of their free trading platform and had a waiting list before they even built it. It was clear that their idea was so good that it appealed to everyone and from there, it becomes an execution challenge, and those kinds of challenges are easier to meet than finding a perfect product market fit.

What do you know now that you wish you had known earlier on?
I know now the power of social media. When it first emerged as a next-generation opportunity, it just

didn't make sense to me and that may be my Baby Boomer orientation. Clearly, Facebook and other social media platforms like Twitter saw the power of creating communities early on and the impact it would have by getting more people's voices into the world. It is incredible what has come out of that and in many ways has become a key driver of our society. I'm not sure that it is all good, but I didn't see that coming.

Would you please create a personal quote that captures important wisdom about participating in early stage investing?
Here I will quote my father who said to me when I was nine years old and being bullied at summer camp –

"You just gotta learn to roll with the punches."

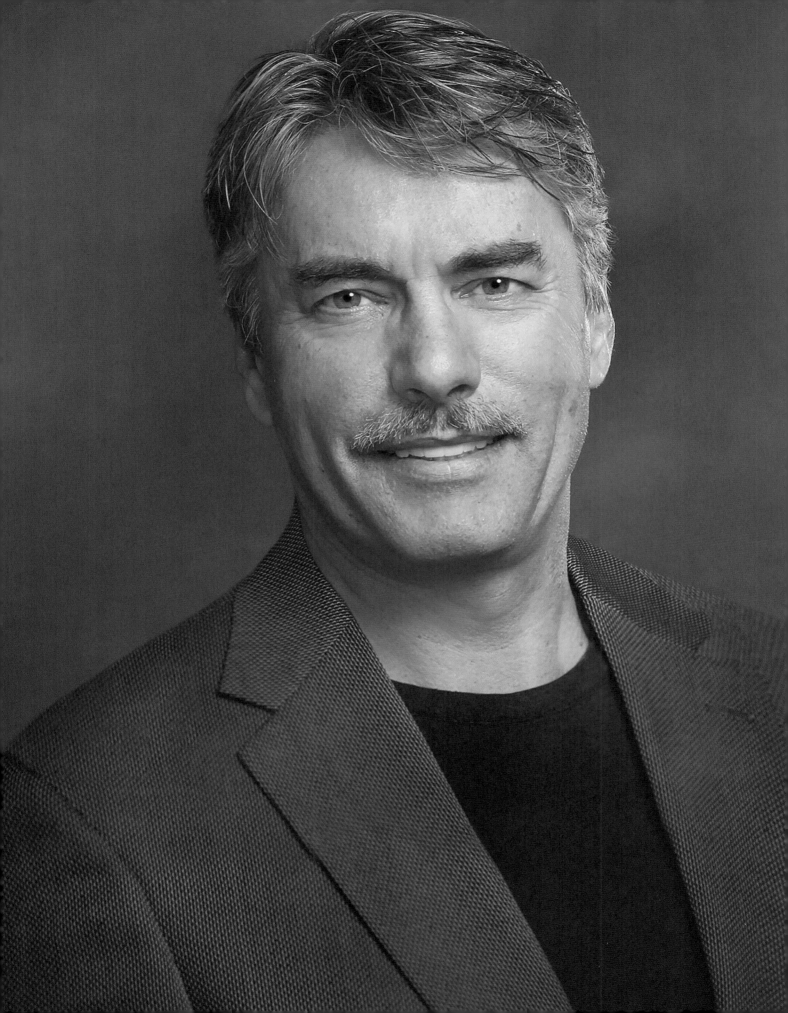

> *"Pursue deals in no more than three proven growth markets and find the promising companies in those markets.*

Robert Canepa

Angel Investor

Keiretsu Forum, Keiretsu Capital, Fifth Era, Blockchain Coinvestors

What did you want to be as a child?
Become a businessman in manufacturing.

What was your first career?
Field Sales Engineer, Texas Instruments, Santa Clara, California.

What other careers have you had before your current one?
Co-founder and President of Fine Pitch Technology, Inc. San Jose, California.

Why do you choose to focus on early stage technology investing?
I wanted to participate in the potential upside of new technology trends without doing the heavy lifting.

When your investing is not going well, what do you do or say to yourself to keep motivated?
I invest only high risk capital. I can now review the original opportunity and find my misstep(s), without feeling too much pain. Now go find the next one, and don't screw it up this time!

What do you love most about what you do now?
Learning of many new emerging markets, and some old markets being re-imagined. Hearing the stories of many CEO's. Their dream, vision, struggle, current challenges and most importantly, why they need your capital.

What dreams do you have for the next 10 years?
That every single one of my portfolio companies has a successful exit. And yes, it is just that, a dream.

What was your first technology investment and what happened?
Advanced motion technologies using nanotechnology. It failed.

Can you share a story of one of your best investments, and why it went so well?
In 1986 I invested everything I had into a small tech manufacturing line in my garage in Milpitas CA. I was out of cash, no furniture in the house, using roommates to pay my mortgage, and my employees, unbeknownst to me, were taking work home at night on their own time to help-out. To meet payroll, I would sit quietly and wait in the lobby of my customers until they cut me a check. By 1988 I was out of the garage, had taken in two partners, and we expanded rapidly in San Jose, CA. We then put a plant in Mountain view, CA, and then a plant in Santa Ana, CA. In 1993 we purchased a building in San Jose for our main plant and headquarters. At our peak, we had 200 employees. All this growth was funded by our customers allowing us high margins due to our technology, high level of quality, and service. At this point we started getting noticed, and a New York investment bank offered us a position in a roll-up, which we ultimately declined, but we were now in-play. An offer to be acquired came soon after, which was flattering, but we still enjoyed our independence and we again declined an advance. We then hired a prominent San Francisco based investment bank to find a merger partner. They found a $1B+ multinational public company, but the initial talks failed to bring us to an agreement. A year later I rekindled the relationship and a merger was then consummated at a high valuation in March 1996. Post-acquisition we expanded with a fourth plant near Boston, MA. By then my time was over, we made an exit agreement and I left the company in 1997. The dot-com boom was just emerging and the public stock I received from the merger achieved a 10x rise in value. This investment was successful because we stayed laser focused on our niche, we held our costs down and maximized our value (margins), and we had a good feel for when it was time to exit.

In your opinion, what is so special about Silicon Valley?

For many years Silicon Valley was known as the best place to participate and compete in the high technology arenas of the future. If a person could compete and win in the San Jose area, they could win anywhere! Personally, I felt I had no place else to go. In 1982 I sacrificed my future with the company that hired me out of college just to get to Silicon Valley. Once there, I was convinced I would find a profitable niche in technology manufacturing. At that time, no other place in the world had the high concentration of diverse and complex semiconductor designs. Many engineering firms were on aggressive design schedules that gave us the opportunity to match high technology and accommodating schedules for high margin manufacturing.

Who have been role models you most admire?

My Dad!

What do you believe are the most important traits of a technology entrepreneur?

A burning desire to succeed. Deep market knowledge and a very keen sense of how to make money in that current or near future market. Ability to adapt when you are wrong, and you will be. Surrounds oneself with profit-skilled partners, investors and advisors. Hires people with a clear service bias, and track records that match company goals. Work ethic; shows dedication and commitment that will inspire the team to excel.

If you could tell a technology entrepreneur just one thing, what would it be?

Build your business in a market that already exists and create margins that fund your growth. If you can't do that, and need to bring in investors, have a clear path to near term profitability. Resist attempting to change the world by spending investor capital and 8 years to create your new idea.

Would you please create a personal quote that captures important wisdom about participating in early stage investing?

"Pursue deals in no more than three proven growth markets and find the promising companies in those markets. If you are going to get yourself in position to find and fund these companies, and achieve a successful exit, you are going to have to make it a full time job."

High risk-tolerance and having the ability to fly blind. That is, to be able to accept – and be "comfortable" with – a high level of uncertainty.

"There are a million reasons to say no – to get to yes, successful early stage investing requires judgment based on diligence, intellectual rigor, and being able to be rationally optimistic."

Asheem Chandna

Venture Capital Partner

Greylock Partners

What did you want to be as a child?
Inventor.

What was your first career?
Product Manager.

What other careers have you had before your current one?
Product management, marketing and business development roles in high-growth technology companies.

Why do you choose to focus on early stage technology investing?
I understand the entrepreneur's journey, having been down that road as a former product executive. I love to dig in early to help founders think through key foundational decisions and help with company building.

When your investing is not going well, what do you do or say to yourself to keep motivated?
Failure is part of the journey. One needs to be willing to take risk and remember that success is not always 'up and to the right.'

What do you love most about what you do now?
We are living in a time of unprecedented change and opportunity. The ability to partner with very talented folks, help create and build important new companies.

What dreams do you have for the next 10 years?
Technology advancements help increase productivity, expand opportunity and help make the world a better place. Continue to partner with entrepreneurs and teams to help build important companies that matter, and that one can be proud of.

If you could change one thing about the world we live in, what would it be?
Equality of opportunity globally, irrespective of gender, race, color, citizenship, social strata, or where one is born.

Can you share a story of one of your best investments, and why it went so well?
In 2005, founder Nir Zuk set out on a mission: disrupt the network security status quo and re-invent the firewall. Greylock, along with Sequoia Capital, partnered with Palo Alto Networks at the very beginning with a $250K seed check. Nir started from a desk at the Greylock offices. Nir was methodical about picking a market that was large and we talked with 40+ customers before formulating the concept. While still at investigation stage, he wrote a detailed marketing data sheet for the 1.0 product, to further convince himself that a customer would have interest in purchasing the potential features/benefits relative to incumbent offerings. Most founders won't go through this level of intellectual rigor. Nir knew his superpower was to be the CTO and we recruited three co-founders, including VP engineering and CEO. The company then focused on building a very strong early product team and iterated closely with customers on delivering a superior 1.0 offering. Over the following years, the business grew rapidly and eventually went public. Post-IPO, PANW has grown into the world's largest cybersecurity company. Some lessons for success:

- Targeting a large market.
- Extreme customer focus.
- Building a stage appropriate team.
- Growing as fast as possible, while maintaining rigor on increasing operating leverage.

Our original investments were written at a $2M and $15.5M company valuation, the company went public at $3B+, and the market cap today is $24B+. The business currently has 8,000+ employees, 70,000+ customers, is $4B+ in billings run-rate, and generates ~$1B of annual free cash flow.

In your opinion, what is so special about Silicon Valley?

Silicon Valley is a special place – based on a unique combination of location and weather; world-class universities, people and talent; cultural diversity; risk-taking mindset and broad access to capital. There are places around the world that have some of the above attributes – but almost none that have all of these. Silicon Valley is also a mindset – where people are willing to meet each other, be helpful across traditional company boundaries, and have a "pay it forward" attitude and culture.

What is the best advice you ever received about how to succeed in Silicon Valley?

A professor in graduate school gave me some very valuable advice – start at a large company to understand excellence at scale, but then quickly move to a smaller company to participate in growth. I was fortunate to start at AT&T Bell Labs in New Jersey; and then three years later moved to Silicon Valley and joined SynOptics, a high-growth startup networking company. Three year later I moved again, and joined a startup that had just raised venture financing as an early employee.

What do you believe are the most important traits of an early stage technology investor?

Early stage investing requires a combination of an investor mindset, coupled with company building skills. The best investors are highly competitive, driven, intellectually honest, rigorous and independent thinkers. They earn the trust of entrepreneurs and push for excellence while also caring deeply.

If you could tell a technology entrepreneur just one thing, what would it be?

Startup journeys are hard but also deeply rewarding. Focus on the customer, recruit the very best team, iterate fast and be persistent.

Would you please create a personal quote that captures important wisdom about participating in early stage investing?

"Early stage investing requires the ability to intersect, assess and win the privilege of partnering with exceptional founders in important markets. The best founders can communicate precisely about a problem and apply rigorous thinking to create an original solution. One is generally investing in a thesis around the team, a north star in markets and envisioning what can be built. There are usually no business metrics and the glass only slightly full. There are a million reasons to say no – to get to yes, successful early stage investing requires judgment based on diligence, intellectual rigor, and being able to be rationally optimistic."

Silicon Valley is not as large as people think it is, and it's really built around a strong sense of community. The best advice I received is to prioritize personal relationships and not to view them as transactional. Always pay it forward.

"Failure does not define your next opportunity."

Gary Cookhorn

Venture Capital Partner

Health2047 Capital Partners

What did you want to be as a child?

Arsonist and Fireman. I loved bonfires (in the UK, it is an annual ritual for kids to celebrate Guy Fawkes night, November 5, with a bonfire). But, I was also thrilled with the idea of racing around town in a big red vehicle with brightly flashing colored lights, siren blaring and a powerful water cannon. (I don't think it took my pediatrician too long to diagnose ADHD).

What was your first career?

I trained in the UK as a Chartered Accountant/ Auditor.

What other careers have you had before your current one?

Worked in finance at UN and World Bank, and later, for different entities as a chief-of-staff, project manager and marketer.

Why do you choose to focus on early stage technology investing?

To help get a seat at the table when decisions are being made.

When your investing is not going well, what do you do or say to yourself to keep motivated?

Don't look down!

What do you love most about what you do now?

The positive impact that the companies and entrepreneurs invested in can have on real people's lives.

What dreams do you have for the next 10 years?

To help make the (highly dysfunctional) US healthcare system work better for all participants – especially patients and physicians.

What was your first technology investment and what happened?

I was an early investor in Bitcoin - made a handsome multiple of invested capital very quickly, then took most of my money off the table (thinking my luck would soon run-out). In retrospect, I sold way too early.

Can you share a story of one of your best investments, and why it went so well?

On a return on investment basis, the investment above was great. I had invested during a period when the price of Bitcoin was wildly volatile. (Little did I suspect that the high volatility would be a fairly constant feature).

Can you share a story of one of your worst investments and why it went so badly?

My worst investment was in German residential real estate (clearly, non-tech). Directionally, it was a good bet – German residential real estate prices climbed as expected. But, the vehicle I had invested in was badly structured from the financial point of view (too much short-term debt) and I lost 97% of my investment!

In your opinion, what is so special about Silicon Valley?

The network effect: Silicon Valley is an entire ecosystem of high-quality talent - entrepreneurs and startups, competitive sources of flexible capital, soft infrastructure and service providers (law firms, accountants, etc.), universities (Stanford, Berkeley, etc.), etc.

Why do you think the innovation economy/ Silicon Valley has a poor track record on diversity and inclusion? What needs to change?

I think the premise of the question is only partially true: There is some diversity in Silicon Valley. For instance, many VC's and founder/entrepreneurs, are of Asian origin – from the Indian sub-continent and China, in particular. However, what is true is that there are far too few women and African-American's, Latino's, etc. Not sure why this is the case. Bias is likely a factor. Without attaining critical mass and a mutual-support network, this issue will remain difficult. But, greater opportunities for STEM training may also help.

There is a lot more focus now by investors, employees and customers on companies that do good (and do no harm) as well as make money. How do you think about this topic in the context of your investing?

Agreed. When it comes to 'doing good', many like to talk-the-talk. But, few can walk-the-walk - and, sadly, when they do, they seldom deliver returns. Investing in healthcare is particularly suited to achieving good returns while doing good. It is a large sector of the economy (approx. $3.5tr, 20% of GDP) and highly inefficient (waste, fraud and abuse is estimated to account for 30% of the cost). Therein lies the opportunity. But, it is also important to hold to account portfolio companies to adhere to ESG principles.

Who have been role models you most admire?

No real role models. But, I admire Eric Whittaker (Zing Health) as an entrepreneur, Pete Briger (Fortress Investment Group) and my colleague, Dr Doug Given (Health2047 Capital Partners), as investors.

What is the best advice you ever received about how to succeed in Silicon Valley?

Don't make emotional decisions.

What do you believe are the most important traits of a technology entrepreneur?

High risk-tolerance and having the ability to fly blind. That is, to be able to accept - and be "comfortable" with - a high level on uncertainty.

What do you believe are the most important traits of an early stage technology investor?

To be unafraid of the dark. The fact that they don't have (good) answers to many questions does not deter them. Granted, sometimes this is because they have not adequately considered the right questions or risks. But, usually, it is their persistence - or deep-seated belief in themselves and conviction in their ideas – that sees them through.

If you could tell a technology entrepreneur just one thing, what would it be?

Things are rarely as bad as they seem, but seldom as good as they appear!

What do you know now that you wish you had known earlier on?

No one really knows anything – including ourselves! If you're doing something new, you'll probably soon be the "expert" in your specific field. Sadly, you're likely learning the hard way, by making mistakes – lots of them – and it is painful. Besides, it takes a long time to train an apprentice on Sand Hill Road. And, paradoxically, the ground is always shifting – what you learned yesterday is often of little value tomorrow. In this environment, it is difficult to acquire a deep reservoir of knowledge.

Would you please create a personal quote that captures important wisdom about participating in early stage investing?

Here are three, though I'm sure none are original:

1. *"Failure does not define your next opportunity."*

2. *"There are only two kinds of entrepreneurs: those who have failed and those who will fail."*

3. *"In the context of great ideas: Over-confidence kills, diffidence gives birth to the stillborn."*

The learning...I learn when I invest; I learn when I teach; and I learn when I mentor.

"The future is up to you. Anything is possible."

Tim Draper

Venture Capital Partner

Draper Associates, Draper Venture Network, Draper University

What did you want to be as a child?
Baseball player.

What was your first career?
Entrepreneur/Marketing for HP.

What other careers have you had before your current one?
Investment Banking with Alex. Brown & Sons. Venture Capital with Draper Fisher Jurvetson.

Why do you choose to focus on early stage technology investing?
Because it can have a lasting effect on the world. Because entrepreneurs make an extraordinary impact. Because it inspires HOPE!

When your investing is not going well, what do you do or say to yourself to keep motivated?
It is just a matter of time before the customers see the light.

What do you love most about what you do now?
Meeting people who help me see the future through their visionary eyes. Seeing the great successes become a part of pop culture.

What dreams do you have for the next 10 years?
I look to a world that is decentralized, to currency like Bitcoin that is global, open and transparent, to health care that uses data to drive diagnoses and treatment. I expect governments will have to compete for us, as we become more mobile and nomadic. I envision some of the biggest industries in the world will be transformed over the next ten years because of AI, Bitcoin, the blockchain, smart contracts and surveillance.

If you could change one thing about the world we live in, what would it be?
I would eliminate dictators, and let freedom ring. I would allow more flexibility to startups to innovate without regulatory friction for 5-6 years. I would encourage governments to compete with each other for us, providing service that matches the cost (taxes) and privatize wherever they could.

What was your first technology investment and what happened?
I came out of the blocks with three companies. One was Home Security Center, a retailer that provided things like alarms and cameras that failed almost immediately. One was Parenting Magazine, which was sold to Time, Inc for 3 times on my money in 6 months, and one was Parametric Technology, a 3-D CAD company that today is the largest software company in the NorthEast.

Can you share a story of one of your best investments, and why it went so well?
One of my best early investments was Hotmail. It was founded by two 26-year-olds, Sabeer Bhatia and Jack Smith who were building web-based email. I suggested that they put a message at the bottom of each email advertising the service, they did it, and it spread to 11 million users in 18 months. The success kept costing us money, as we needed to keep buying servers, so we had to find a partner. Microsoft bought it for (as reported) $400 million in MSFT stock, and to this date, Bill Gates says that Hotmail was the best acquisition he has ever made.

Can you share a story of one of your worst investments and why it went so badly?
The worst investments I have made are the ones that keep looking like they are going to make it with just a little more money, and don't. They cost me the most. The worst decisions I have made are when I didn't invest in Netflix or LinkedIn.

In your opinion, what is so special about Silicon Valley?
Silicon Valley has evolved over the 35 years I have been working here. It started with engineers selling interesting equipment to other engineers. Then there were a few people who were willing to invest money into these engineers. Then my grandfather, William H Draper Jr. became the first of many Silicon Valley VCs. The VCs brought opportunity to engineers and entrepreneurs, and they flourished.

Then the other service providers joined in to help out. Today, the systems are all in place, and the spirit of entrepreneurship has prevailed, and now almost everyone in the Silicon Valley looks at a product or service and thinks about how they might do it better or better delight the customer. You rarely see that kind of spirit anywhere else around the world, but there are starting to be other hotbeds, and I think the world will be amazing as this spirit spreads.

Why do you think the innovation economy/ Silicon Valley has a poor track record on diversity and inclusion? What needs to change?

I disagree with the premise. About 80% of the companies I back have an immigrant in the team. We back people of all religions, races, sexes and nationalities. We fund a disproportionate number of women, because there are so few female entrepreneurs willing to start businesses, that the few we do see are often extraordinary, and we look for those extraordinary individuals who can impact the world with their visions, their innovations, and their technologies. My daughter, Jesse Draper with Halogen Ventures won't fund a company unless it has a female in the founding team.

There is a lot more focus now by investors, employees and customers on companies that do good (and do no harm) as well as make money. How do you think about this topic in the context of your investing?

Whenever I look at a new startup, instead of (or in addition to) looking at all the things that can go wrong, I ask myself, "What if it works?" Then I decide whether I am on board on the world that this entrepreneur is envisioning. And if I love it, I will invest. Sometimes companies will lose their way, but in general, entrepreneurs start companies to make the world a better place. From my POV, the people that have been making life worse are the large companies, whose lobbyists in conjunction with the government officials that create regulations, create rules that favor big business and big government and get in the way of this positive progress that the entrepreneurs are creating for us. I would be careful with this thought process that some politician, government official, religion, or media source knows better what is "good" than the individual "good" people who live and work to make their lives better under the laws of the land.

Who have been role models you most admire?

Steve Jobs and Elon Musk. These two extraordinary people are willing to take a lot of heat in the process of building something that they know in their hearts is going to be good for humanity. Jobs used to fire people who wouldn't have the spirit of accomplishing the impossible (one button iPod for example). Elon tells us that SpaceX will take us to Mars, while all the cynics and naysayers attack and pick on him while he hires the best and the brightest engineers to accomplish this mission that may just save humanity.

What is the best advice you ever received about how to succeed in Silicon Valley?

Connect with people. Everyone can teach you something. It doesn't matter who is buying and who is selling. The human connection matters.

What do you believe are the most important traits of a technology entrepreneur?

The knowledge that if they work hard, hire good people and make the right moves, they have a chance to create something extraordinary that didn't exist before they got started.

What do you believe are the most important traits of an early stage technology investor?

There are good tech investors that come from a variety of different backgrounds. My approach is to try to figure out what the world will look like in the future, but to be willing to be flexible with that vision as I meet new entrepreneurs. I also like to go after markets that I know are providing poor service for a high price, (e.g. banking, insurance, health care, real estate, and government).

If you could tell a technology entrepreneur just one thing, what would it be?

Do what you say you are going to do. As a leader, you will also need to be a model for everyone who works for you, so work the way you want them to work.

What do you know now that you wish you had known earlier on?

I like the Henry Ford quote that says, *"If you think you can or if you think you can't, you are right."*

Would you please create a personal quote that captures important wisdom about participating in early stage investing?

"The future is up to you. Anything is possible."

Resilience. Occasionally entrepreneurs get fortunate enough to enjoy a quick turn and exit fast, but in general, the startup game is a slog. You're attempting to defy the odds, break down barriers and succeed where others have no doubt tried and failed before. Competitors will push you to fail. Without grit, you're going to tap out before you reach the finish line.

> *"Invest in the people who give you reason to believe they have the potential to build something truly special, then do everything in your power to help them achieve that potential."*

Dan Elitzer

Venture Capital Partner

IDEO CoLab Ventures

What did you want to be as a child?
Farmer.

What was your first career?
Business consulting for nonprofits and social enterprises.

What other careers have you had before your current one?
Finance and operations for microfinance and international development organizations.

Why do you choose to focus on early stage technology investing?
I was already spending so many hours focused on early stage tech, it seemed like the best way to align my vocation with my time.

When your investing is not going well, what do you do or say to yourself to keep motivated?
Luck plays an enormous role in individual outcomes. However, process can be controlled, and luck favors the prepared. Focus relentlessly on process.

What do you love most about what you do now?
Having an excuse to constantly meet, learn from, and be inspired by people much smarter than myself.

What dreams do you have for the next 10 years?
Professionally, I hope to play a role in supporting the success of a venture that positively impacts the lives of more than a billion people. I believe emerging technology in general—and crypto and open financial technology, in particular— will have a massive impact on the world and I am excited to help bring that potential to fruition. Personally, I look forward to enjoying and deepening my relationships with my wife, my children, and other family members, as well as close friends.

If you could change one thing about the world we live in, what would it be?
I'd love empathy to be much more common. So many of our current tools and social systems encourage conflict and extreme responses. If we can find ways to remember that we're all human and trying to build better lives for ourselves and our communities, perhaps we'll find more opportunities to work together rather than tearing each other down.

What was your first technology investment and what happened?
My first technology investment was in Bitcoin in 2013. I bought my first Bitcoin at $90 and sold it couple months later at $120 and was thrilled. It took me a little while to realize I should be investing over long time horizons rather than trying to time trades for short-term gains. The experience of putting my own money on the line—as small as the amount was—sharpened my desire to learn more about Bitcoin and the technology behind it. As I started to travel down the rabbit hole, I began my MBA at MIT Sloan and founded the MIT Bitcoin Club. Prior to that, I'd been working for nonprofits and was now taking on significant grad school debt, so my investment was mostly my time. I probably spent more time focused on Bitcoin and building up the MIT ecosystem around it than I spent on my classes or the typical MBA networking activities.

That investment really paid off. I got recruited out of school by IDEO to help start a collaborative R&D studio focused on blockchain technology. Over the past five years there, I've been able to work with incredible colleagues and partners. Two years ago, we launched a venture fund focused on distributed web technology, a term we use to talk about the broader space around Bitcoin, blockchains, and crypto. That initial investment in Bitcoin set me on a path to find my calling as a venture investor. What better return could I ask for?

Can you share a story of one of your best investments, and why it went so well?
I'm still a bit early in my investing career to have a completely proven winner with a full exit. That said, my best investments have tended to be ones where by the end of the first meeting, I'm strategizing with the founders about how to make sure they keep heading

in the right direction. Successful founders tend to move fast and with a strong sense of urgency - when I find myself drawn to match their pace, that's usually a good sign.

Can you share a story of one of your worst investments and why it went so badly?

My worst investment thus far was one where I got so enamored with the idea that I brushed aside my initial concerns that the founders weren't the right ones to pursue it. As it played out, I saw them make mistake after mistake. I spent a lot of time with the founders trying to coach them and handle some of their biggest problems, but I really should have thrown in the towel earlier. It ended up being not just a poor financial investment but also a poor investment of my time and energy. The biggest lesson I took away from this experience was to never invest in a company where I believe I am better equipped to run it than the founder. I look at hundreds of companies per year, but I become an expert in very few. I also just don't have the amount of time and mental cycles to dedicate to any one company that can hope to match the founder. Despite that, there's a strong desire to try to help when you see a company struggling – but an investor can't be the one driving a business forward.

In your opinion, what is so special about Silicon Valley?

There are two mental orientations that are common in Silicon Valley, but much rarer in the rest of the world. The first is a belief that it is genuinely possible for a talented individual or small team to build something that will eventually touch the lives of millions or even billions of people. Your dreams can be your limits, so set them high. If you want to build a business on a global scale though, you can't be conservative in how you go about getting there. That brings us to the second mental orientation of Silicon Valley: failure is accepted as part of the process. This applies to both decisions made while operating a particular startup, as well as pulling up and looking at the startup as part of the broader innovation and venture ecosystem. Embracing the possibility or even likelihood of absolute failure is key to unlocking the possibility of extreme success scenarios.

While I view these frameworks as being a large part of what makes Silicon Valley special, there's no question that they can sometimes be taken to extremes. It can be easy to get too comfortable this social bubble and forget that most people in the world don't think this way, or delude ourselves into thinking there are no dangers or broader externalities that can result. For Silicon Valley to remain a shining light of innovation in the world, we're going to need to find ways to reconcile the best of what's been working with a need to welcome participation from other communities around the world.

Why do you think the innovation economy/ Silicon Valley has a poor track record on diversity and inclusion? What needs to change?

In Silicon Valley, everyone is trying to make decisions and move as quickly as possible. When you're trying to move quickly and on instinct, you're naturally going to end up drawn to people who are similar to yourself or to people you've worked with in the past. The diversity problem that arises from this is exacerbated by the fact that those most able to take the risk of starting a company tend to come from privileged backgrounds, so the status quo tends to perpetuate as a result. I'm not saying the impulse to construct teams out of your close personal network is necessarily wrong early on, but there are tradeoffs. If this dynamic is not examined and steps are not taken to intentionally cultivate diversity in the first 20 or so members of a team, it gets really hard to course-correct later. There is plenty of literature showing the benefits of having diverse and inclusive teams. Getting there is not easy or automatic, but it's worth the effort. Hopefully we're all building companies that are a microcosm of the world we want to live in, so we all need to do our part to better reflect that.

There is a lot more focus now by investors, employees and customers on companies that do good (and do no harm) as well as make money. How do you think about this topic in the context of your investing?

I honestly believe a lot of decisions that get framed as a tradeoff between doing good and making money aren't really about trading one for the other; it's mostly a question of time horizons. It's possible to be short-term oriented and value extractive over months, quarters, or even years, but when you think in terms of maximizing profit over decades or longer, a lot of things no longer seem like tradeoffs. My investing has focused on financial technology, almost exclusively on opportunities related to Bitcoin, crypto assets, and open financial systems. Anyone who has spent more than a few minutes looking at crypto knows that there are plenty of opportunities to make (or lose) vast sums of money over short time horizons. In 2017 in particular, during the ICO (initial coin offering) boom, many supposedly long-term venture investors would buy into private rounds, wait a month or two for the token to launch and get bid up, then quickly take profits and move on to the next thing. That's not how I want to invest, and it's not even about the ethics of those particular behaviors. When I invest, I'm committing to go on a journey with founders over many years. If I don't believe that a future where that company is successful is one that leads to a better world, why would I want to go on that journey?

What is the best advice you ever received about how to succeed in Silicon Valley?

Don't take "no" for an answer. There is almost always another way if your goal is important enough. This goes for life, not just Silicon Valley.

What do you believe are the most important traits of a technology entrepreneur?

Speed of execution, the ability to see clearly and the ability to adapt. How fast an entrepreneur charges towards her goals with purpose is probably the most important signaling factor for me. When moving fast, it's also important to be able to see things accurately for what they are. While a certain level of reality distortion field is very helpful for entrepreneurs to make sales and recruit a team, the best tend to be able to see reality clearly for themselves, which allow them to adapt as necessary.

What do you know now that you wish you had known earlier on?

When assessing a founder or an investment opportunity, first impressions are important. However, even more important is how much they improve from a first meeting to a second (or third, or fourth) meeting. You don't invest in founders for who they are or what they're capable of today; you invest in them for what they're capable of becoming and accomplishing over the months and years ahead. Drawing that trend line accurately is much easier when you can plot multiple dots.

Would you please create a personal quote that captures important wisdom about participating in early stage investing?

"As enamored as we get with ideas, early stage investing is largely about people. Invest in the people who give you reason to believe they have the potential to build something truly special, then do everything in your power to help them achieve that potential."

"When opportunity knocks, answer the door!"

Alan Fisher

Angel Investor and Serial Entrepreneur

Sand Hill Angels

What did you want to be as a child?
Engineer.

What was your first career?
Software engineer.

What other careers have you had before your current one?
Venture capitalist.

Why do you choose to focus on early stage technology investing?
I'm a serial entrepreneur myself and enjoy helping other entrepreneurs. Plus, I'm always learning something new and am constantly amazed by the innovation occurring all around us.

When your investing is not going well, what do you do or say to yourself to keep motivated?
Angel investing is a hit record business, so portfolio diversification is key. Invest in high quality companies with high quality teams, and your portfolio will turn out well in the end.

What do you love most about what you do now?
I'm always learning something new about areas I'd never thought about or considered.

What dreams do you have for the next 10 years?
All my portfolio companies go public! On a more serious note, as a member of Sand Hill Angels, one of Silicon Valley's large angel investing groups, I see a lot of pitches and am continually impressed by how much innovation is going on, in all areas. It's not just software and Internet anymore; it's everything and everywhere.

If you could change one thing about the world we live in, what would it be?
The general level of financial and business literacy among the general public is appalling. If I could change one thing, I'd require financial and business in our schools.

What was your first technology investment and what happened?
I had the good fortune to invest in Microsoft, Oracle, Lotus Development and others shortly after their IPOs. My learning there is great companies have volatile stock prices as they grow.

My early angel investments were in companies that I co-founded, two of which went public. Other angel investments where I wasn't a co-founder had mixed results; some went to zero and others had awesome returns, proving that angel investing is a hit record business so it's best to build a diversified portfolio.

Can you share a story of one of your best investments, and why it went so well?
A young colleague and serial entrepreneur who I greatly admired was raising money for his new company. We had a few preliminary conversations about me investing and I tentatively agreed, though his business concept was spongy and rapidly evolving. He came to my office one day anxious to get my check. I said, "Do you need it to make payroll?" and he sheepishly said yes. So, I wrote him a large check and two years later that investment returned 15X when he sold out to a public company. To this day, I still don't know what his company actually did!

Can you share a story of one of your worst investments and why it went so badly?
I invested in a startup that was first-to-market in DVD rental kiosks, two years before Redbox was born. After investing, it became apparent the two co-founders couldn't get along and were purposely avoiding each other to the extent they couldn't be in the office at the same time. That, coupled with technical difficulties with the kiosks that could have been overcome, was too much for the small team of six to handle. I took over as Chairman, but there was not much I could do at that point and the company and my investment went to zero.

In your opinion, what is so special about Silicon Valley?
Silicon Valley has many entrepreneurs who have had successful exits and are willing to recycle their capital

into other high-risk startups. I don't believe that concentration of high-risk capital exists anywhere else in the world. Of course, pools of high-risk capital will form over time in other locations outside of Silicon Valley as successful exits occur, but for now, it makes Silicon Valley unique.

There is a lot more focus now by investors, employees and customers on companies that do good (and do no harm) as well as make money. How do you think about this topic in the context of your investing?

As an angel investor, I see an incredible array of innovative companies, all singularly focused on making the world a better place, one niche at a time. I am blessed to live and invest in an environment where this is systemic.

Who have been role models you most admire?

Jerry Kaplan, Chris MacAskill

What is the best advice you ever received about how to succeed in Silicon Valley?

Cash is more important than your mother! A successful serial entrepreneur told me this when I was launching my first company.

What do you believe are the most important traits of a technology entrepreneur?

Adaptability and the ability to pivot quickly upon encountering new information and insights about your customers and market.

The ability to communicate crisply and clearly to your customers, employees and investors.

What do you believe are the most important traits of an early stage technology investor?

Inquisitive – willing to learn about new things

Helpful – willing to actively help entrepreneurs

Risk tolerant – because 90% of your portfolio will (mostly) fail.

If you could tell a technology entrepreneur just one thing, what would it be?

Cash is more important than your mother! A successful serial entrepreneur told me this when I was launching my first company.

What do you know now that you wish you had known earlier on?

Investors invest in businesses, not markets. Customer traction is vital, and you should engage with customers the moment your business moves beyond the cocktail napkin stage.

Would you please create a personal quote that captures important wisdom about participating in early stage investing?

I'll give you several to choose from:

1. *"Water your winners and pull your weeds!"* (re-invest in winning businesses)

2. *"Time kills all deals."*

3. *"When opportunity knocks, answer the door!"*

4. *"Cash is more important than your mother!"*

The level of curiosity is obscene. It does not matter the industry, from funerals to farming, crypto to camping there are entrepreneurs trying to work out how to innovate to build a business. The more obscure the more interesting. I love it.

"Take a big picture – reflect on the ingenuity of the world and know that we are all, as a community, full of amazing capacity to solve and innovate."

Gamiel Gran

Early Stage Venture Capital –
Business Development

Mayfield

What did you want to be as a child?
Doctor – Emergency Room.

What was your first career?
Graduated high school early and took a job in the plant nursery business. I had had my own landscaping business in high school and was trying to earn cash to pay my own way for college. I took a job at a local nursery, and the owner took me under his wing and encouraged me to take State test to become California Certified Nurseryman. I did this before going to Cal Berkeley.

What other careers have you had before your current one?
Always been in Hi Tech. I interned at IBM while going to college and started at IBM after graduating. Went from there to Oracle, BEA, plus a few startups before entering the Venture Capital world.

Why do you choose to focus on early stage technology investing?
Early stage is the most exciting, fun stage of business building. Highly enthusiastic entrepreneurs, who want to change the world is always inspiring to be around. Most of them are humbly seeking as much support as they can get to help build their business. This makes the role I provide a real win-win, whereby I help startup teams reach early customer adoption, i.e., ICP (ideal customer profile) and PMF (product market fit).

When your investing is not going well, what do you do or say to yourself to keep motivated?
Just simply remember that not all investments will work…in fact many will fail. Remain focused on the broader investment thesis and look for gaps in market, team and product fit to refine or pivot the deal.

What do you love most about what you do now?
I am so grateful to work with smart, enthusiastic entrepreneurs who are on a mission to solve big problems. It's exciting to be part of their big vision. Helping entrepreneurs reach their goals generally means improving the world in some way. There's a lot of hard work, but balanced with how passionate they are, and when they achieve success, how rewarding it can all be – and how the rewards are for the customer, entrepreneur and the investors.

What dreams do you have for the next 10 years?
I am deeply inspired by how the world is evolving around more important issues of diversity, human and planet health. I believe we are at the forefront of truly amazing innovations across human, animal and plant biology that will change the world for the better. The first era of technology innovations were with semiconductors, networking, the Internet, mobile and the cloud. The next era will unleash even bigger dreams. Technology and biology are now joining in the creation of a new renaissance of innovation. We are opening new insights at an unprecedented scale as we now have the ability to do full genome sequencing for plants, animals, humans plus materials. Technology of vast cloud networks, advanced data science, and networks make the discovery process simply more accessible. Perhaps we are at the next stage of technology, with first round of the past 30 years opening the doors to a more impactful era for the next 100 plus years. The imagination of entrepreneurs plus the support from thoughtful long-term investors will combine to shatter the problems that previously were unsolvable.

If you could change one thing about the world we live in, what would it be?
Human creativity is untapped. But only a small portion of the world is wealthy or educated well enough to have the opportunity to unearth their creativity. I am a strong believer in individual destiny and our natural ability to solve problems. High quality education, for all people, in all parts of the world has the ability to solve racism, diversity issues, women's rights, climate change, costs of healthcare, and so many unfortunate problems that continue to plague so many people. Advances in information

access, online education, and resources for educators are a part of the puzzle. The technology exists today to make education accessible to all young people of the world. Closing the gap of the have's and the have not's on education is how we can change the world.

What was your first technology investment and what happened?

I was part of a founding team during the boom of the dot-com era. It was amazing times, with so much excitement about how the Internet would change the world, and how every small and large business would be forced to rewrite its strategy to survive. Technology platforms were not yet built, but visions were huge. This was more than a subtle disconnect between reality and dreams. Furthering the fuel, the irrationally exuberant stock market was fodder to the retail investors, providing a sense of endless capital. VCs also would talk about 'money being free', and expansions of small businesses were happening at a pace that was out of control for sure. Meanwhile corporate venture teams wanted in on the excitement too and heaped their added investment dollars into the overheated market. So, what happened? We overdrank the Kool Aid! Too much money was raised, with too many expectations, and not enough real technology to support the grand vision. We failed, and sold the remains, while me and many others from the team scattered to safer roles. We spent the next while of time reflecting on what we missed, while arguing with the IRS on stock option 83b valuation tax burdens. Oh, the lessons of the dotcom era. *"Those who cannot remember the past are condemned to **repeat** it."* George Santayana.

Can you share a story of one of your best investments, and why it went so well?

For me it comes down to believing in the team. The idea itself may be raw. The market may also be unclear or un-evolved. But the team who will work tirelessly are where the makings of a great outcome are built. I met 2 founders who were so passionate about changing the world of transportation, they oozed a level of imagination and vision of what could be possible that just sucked you in. You only wanted to help and be part of this dream. But reality of realizing the dream became clear that it would not really happen. So the team, pivoted away from their first execution plan to an entirely new strategy, but never giving up on their big vision to change transportation for the world. What made this a great investment was the team and specifically the clarity of their mission. They offered a self-aware, and humble approach to working through problems. They would take input from many, listen intently to make sure they truly understood, and then they will execute a new plan. This didn't mean they would take the advice as the answer, but rather as input. They would challenge their original plan, thoughtfully with the new input and sometimes take only some of the input.

Founders with a mission, who are self-aware, eager to learn, but also guided by a 'north star vision' are the most enjoyable to entrepreneurs to work with. And in this investment, the results were a great outcome for the founders, investors and their customer. After nearly 10 years there was a successful IPO, and even more important change to global transportation was also realized.

In your opinion, what is so special about Silicon Valley?

It's not special, but we make it so. Of course, the weather is awesome, the geography is magical, and the landscape is accessible. We all know the history of plume orchards turned into semiconductor plants and the circle of wealth created in the chain between investors who meet entrepreneurs who meet investors and repeat the cycle in an ever-expanding cycle of more ideas and more wealth. But why does this only happen here in the Valley? My answer is Diversity. Diversity in ideas, diversity in people, and diversity in the roles that people can take. It seems it genuinely acceptable to be different in the Bay Area, to come from a different background and to leave a big company job for a fledgling startup with no revenue but a big idea. It happens all the time, and we don't just accept it, we celebrate it. The way we go about Celebrating entrepreneurial and cultural diversity combine to make this an amazing place to live.

Why do you think the innovation economy/ Silicon Valley has a poor track record on diversity and inclusion? What needs to change?

Access to education. The entrepreneurs who present ideas are often from the best universities. They are often of Asian (India, China, Korean or Japanese) or Caucasian. Less often do we see Hispanic or African American entrepreneurs. I suggest that we have not provided an equal playing ground for all. As colleges demand even greater SAT and GPA scores, we know that those students whose parents or support groups can provide special tutoring or SAT prep training are more likely to win an admission to their college of choice. Colleges are doing some to improve admissions for lower income, but the beginning of the education track perhaps in K-6 grades is where the impact is most critical.

Women are more involved today than a generation ago. Title XI changed access to sports for young women over 30 years ago, and perhaps signals the length of time it takes to change a multi-generational gap. There's more work to be done, and investors need to push for diversity on the board of directors, on the management teams, and to remove their entrepreneurial bias for certain universities or backgrounds that are most familiar.

There is a lot more focus now by investors, employees and customers on companies that do good (and do no harm) as well as make money. How do you think about this topic in the context of your investing?

We feel strongly that conscious capitalism is important and logical. This means thinking about the social impact of technology, beyond the technology itself. This means creating opportunities for diversity, equity and inclusion. This means bringing an ethical lens to how leaders treat their employees. This means looking at how can innovation simultaneously re-invent trillion-dollar industries while improving our lives and the state of our planet.

Who have been role models you most admire?

I have a number of personal role models who help me stay centered to think of the world we live in with love and kindness – the most important is my wife who always inspires me to stay centered on what's most important. But in a business world there are some masters of inspiration. Bill Gates for applying his wealth and technical prowess to even bigger goals, Steve Jobs for inspiring the brash, beyond-the-lines innovation, and Marc Benioff for caring deeply about creating a lasting impact with early 1% pledge and countless donations to support the underserved.

What is the best advice you ever received about how to succeed in Silicon Valley?

I've been blessed to work with so many smart, and hardworking people here in the Valley. Be bold, Be prepared, and GSD (Get Sh@t Done!) are the pragmatic words of wisdom I've often received. Meritocracy demands great merits, therefore those who are willing to put in the extra time. Go beyond the expectation. Set a personal higher bar than is expected, and then overachieve are what is expected of this high-paced, always-innovating culture that we live in. Balancing this with the effort to help others, mentoring, making that extra effort to be the reference for someone is the way to survive this Valley. Hard work. High Impact. Give Back and Support Your Network. And Take Care of Yourself and Your Family.

What do you believe are the most important traits of a technology entrepreneur?

"Founders with a mission, who are self-aware, eager to learn, but also guided by a 'north star vision' are the most enjoyable to entrepreneurs to work with. And in this investment, the results were a great outcome for the founders, investors and their customer. After nearly 10 years there was a successful IPO, and even more important change to global transportation was also realized.

What do you believe are the most important traits of an early stage technology investor?

We believe in a concept, People First Investing, which supports the success of the entrepreneur, and considers the bigger goals of building a great business that lasts and not just a flip or a quick return. Listening, coaching, consulting are some of the traits, but mostly being a partner to the entrepreneur for whom they can lean in for input not direction.

If you could tell a technology entrepreneur just one thing, what would it be?

Think big. Let your vision of enormous impact, world-changing idea take hold before all the limiting realties seep in. And in doing so you will build a visceral level of contagious enthusiasm that is so attractive to investors and others who will want to join you in helping you create your vision and turn what was merely a dream into a true reality.

What do you know now that you wish you had known earlier on?

Think bigger and execute. There's so much wealth of money, resources, advice and more, here in the Valley that we should be less cautious and more bold and bigger in our ideas.

Would you please create a personal quote that captures important wisdom about participating in early stage investing?

"We are just beginning, and what a wonderful time to be alive! I feel so fortunate most of the time. When I don't, or I fall prey to the minutia of the tactical, I just try to help myself pull back and view the lens more strategically. This gives me solace, peace and a smile."

"Take a big picture – reflect on the ingenuity of the world and know that we are all, as a community, full of amazing capacity to solve and innovate."

"Smile with your heart and share that smile!!'

"Good works, Good people, Good ideas."

Gregory Hamm

Angel Investor and Founder

Keiretsu Forum, Stratelytics LLC and Agni Energy

What did you want to be as a child?
Tom Swift, Jr. Space Adventurer.

What was your first career?
Either construction laborer (first full time job) or civil engineering research for Kaiser Hospitals (first full time job out of college).

What other careers have you had before your current one?
Business consultant, financial analyst for two startups, economist.

Why do you choose to focus on early stage technology investing?
Expect higher returns than less risky investments. It is fun. Meeting so many smart, good, sincere, creative people. Listening to and learning from them.

When your investing is not going well, what do you do or say to yourself to keep motivated?
Remember the odds, each investment is a 1 in 10 bet. I will have a lot of losers, but I am diversified. It's still fun.

What do you love most about what you do now?
I really enjoy building quantitative models that provide new insights. The intense learning that starts new projects. Interacting with bright, young, creative, energetic people.

What dreams do you have for the next 10 years?
Keep working productively. Get engaged in at least one project to stop or ameliorate global warming. Switch from accumulation to dispersion for the public good.

If you could change one thing about the world we live in, what would it be?
"Therefore all things whatsoever ye would that men should do to you, do ye even so to them." King James Bible.

What was your first technology investment and what happened?
2001, drug development with a small core group and extended virtual team. We had a good idea and a talented core team. But in retrospect, I was terribly naïve about our chances of success. I was the third member of the team. The CEO had very unique experience in successful drug development using a small core team and outsourcing the entire drug development program. The CTO was a very respected software developer with new ideas about structuring data for drug development. I had just finished several successful consulting projects on drug development strategy. The plan was to raise tens of millions and assemble a portfolio of promising drugs. I was impressed by the CEO and CTO, but skeptical, because of the size of the raise. We put together a presentation and got an appointment to present to a well-respected VC. The VC paid attention, asked serious questions, and said she would get back to us. I was hooked. I thought this important person is investing her extremely valuable time and taking us seriously, we must have a chance. None of us had been successful or unsuccessful founders before. We did not have a working version or even a mockup of the software. We had no patents, just vague plans for patents. We were asking for a huge amount of money. All these things would raise red flags for me as an investor today. We kept getting on calendars and sometimes even got called back for second presentations. We made 30 or 40 presentations to 25 or 30 different potential funders. It kept feeling like we were close. I don't think we ever were. In the end, we were all tired out. Two Brits with the same idea, a track record of successful investment, and tens of millions already lined up hired our CEO. The CEO and I remained friends, but he never brought me on board.

In your opinion, what is so special about Silicon Valley?
Culture, resources, chance; all played and play a role in making Silicon Valley special. I think that the role of culture tends to be over emphasized and the role of chance under emphasized. I believe that the US

and the West Coast particularly is more willing to try new things and take risks. However, the role of this cultural bias has taken on mythological importance in some accounts of what makes Silicon Valley special. The US and Silicon Valley also has critical resources: wealth, brains, and great weather. Taking risks on a large scale requires wealth and deep capital markets. From 1935 (graduation of Hewlett and Packard from Stanford) until 1972 (founding of Kleiner Perkins), the US undoubtedly had the deepest capital markets in the world. Brains, the Bay Area has two great universities, Stanford and UC Berkeley. Brains are very mobile, no one wants to leave the weather in Silicon Valley, definitely a big impact. Chance, I think Silicon Valley could have happened elsewhere: LA, Boston, Austin, the Research Triangle, etc. The spark happened here. Hewlett-Packard laid the foundation. Shockley created (with others) the semiconductor and invited the founders of Fairchild Semiconductor to Menlo Park. Kleiner and Perkins came from Fairchild and HP respectively to form the first VC firm. Clustering whether it is Staffordshire potteries in early industrial England, steel in Pittsburgh, cars in Detroit, or VC's and tech companies in Silicon Valley is a powerful driver of mutual growth. After the spark, the learning, the spinoffs, the supporting infrastructure all have led to the concentration and expansion of Silicon Valley.

Who have been role models you most admire?

My father. His life was focused on working (not work) and family. If friends or family needed help, he helped, he was always working. One day we were breaking rocks together (a typical father son activity). My mind was wandering but I realized that he was totally focused on the best way to break those rocks. I was amazed.

What do you believe are the most important traits of a technology entrepreneur?

A belief in the value and eventual success of her idea. The power to communicate her ideas clearly to others. Entrepreneurs don't need to be great sales people, they need to be able to sell (explain) one great idea. Entrepreneurs need to create new relationships and network. To grow, an entrepreneur needs advice, investors, and employees. Usually, a large network of people will have contributed in some way to a successful company. A good idea, an idea that is both unique and high value.

What do you know now that you wish you had known earlier on?

Two things. 1) The importance of proof that people will pay money for the product. I have been surprised at how often I have thought there was an obvious market, but the sales have just not been there. Hard proof of market demand is significant. 2) You need to think hard about the next round of funding. Will someone fund that next round that is five to 20 times the current raise? Companies run out of money with ideas that are still solid.

Would you please create a personal quote that captures important wisdom about participating in early stage investing?

"Good works, Good people, Good ideas."

Good works has two meanings. First, that the product provides something of real value to the market. Second, perhaps not related to investment success, I want to feel good about what my investment is accomplishing in the world. Similarly, good people means both a smart and competent team and that the founders are trustworthy. Good ideas, I put a high value on unique approaches to significant problems.

I like to see a burning passion in the entrepreneur for their company to succeed, an enthusiasm for working as a team member and a willingness to listen carefully to advice.

> *"Follow the talent. Figure out who are the real builders, working on real-world problems and the rest will fall into place. Ignore the hype and the showmanship."*

Sam Harrison

Venture Capital Partner

Blockchain.com Ventures

What did you want to be as a child?
I appreciate that this is broad, but I just knew from a very young age that I wanted to be a "businessman" like my father.

What was your first career?
Finance, I worked at Morgan Stanley in London.

What other careers have you had before your current one?
I was a part-time App dev/publisher (iOS) during post-graduate studies, as well as a brief stint in tech sales & investment banking. I then led early stage tech investing at Naspers Ventures.

Why do you choose to focus on early stage technology investing?
Early stage technology investing provides a sneak peak of what the future could and probably will look like. Every day is intellectually stimulating.

When your investing is not going well, what do you do or say to yourself to keep motivated?
The great thing about a diversified portfolio is that there's <u>always</u> something exciting happening at any given point in time. I'm energized by the small wins that companies regularly achieve.

What do you love most about what you do now?
The people I interact with are of the highest caliber. I'm constantly learning about new businesses and also learning from the people that run them.

What dreams do you have for the next 10 years?
I'm choosing to interpret this question in a professional context. As such, I'd love to see mainstream adoption of blockchains and cryptocurrencies, starting with the parts of the world that they could benefit most. It would be great to see securities trade 24/7 and become globally accessible. Large popular virtual worlds with no restrictions on item movement, ownership and trading. Remittances in seconds and for a few basis points, not days and >5% fees. Ticketing marketplaces with guaranteed authenticity and close to zero fees. All can be made possible with blockchains! More personally, I'd love to finally learn a 2nd language.

If you could change one thing about the world we live in, what would it be?
I believe that the world would be a better place if it were closer to a true meritocracy. Today, the world isn't even close to being fair. It matters too much where you're born. The wealth of your parents. Your race, sex and age also matter too much. Talent is evenly spread, and opportunity should be too.

What was your first technology investment and what happened?
The first tech investment I made was buying stock in 2009. I didn't know much, only that they licensed out the IP for the iPhone CPU, but that was enough to get me convinced. The first venture investment I made was in a private social EdTech platform, which has become the second largest educational platform on the planet, measured by web traffic, after Wikipedia. The company has grown to service over 150m monthly unique users (as of 2019) and has raised $45m more subsequent venture funding over two up-rounds. I count the CEO/founder as a personal friend.

Can you share a story of one of your best investments, and why it went so well?
I bought Bitcoin in 2012 and got extremely lucky. I was fascinated by *magic internet money* and it turns out a few other people were too.

Can you share a story of one of your worst investments and why it went so badly?
I've been fortunate enough to make venture investments in over 40 companies. I've been even more fortunate to have only taken one impairment although the laws of venture investing statistically dictate that more must be coming. That empirically poor deal was a small angel investment into a company that I clearly didn't understand. I knew I didn't understand it and I actually still don't. It is in the biotech space, which is an area of keen interest,

but also not my sector of expertise. I didn't make it past A-level biology. I was assured that this company had the highest quality investors and advisors, incredible science and would IPO the year after. Fast forward two years and it is valued at 60% less now, based on the latest funding round. Clear rookie mistake to invest in something I knew nothing about, even if it was a small angel cheque. I was chasing a quick IPO, following the hype, and trusted the pedigree of other investors, rather than doing my own work. Moving fast just to be competitive to get into a "hot" deal is also dangerous. I prefer to develop a great relationship with the executive team and take my time in doing so. If the relationship is strong enough, they'll often ensure you have all the time you need to complete your process.

In your opinion, what is so special about Silicon Valley?

I've lived in 6 countries, so feel reasonably well placed to compare and contrast other cities. My verdict is that Silicon Valley is the most concentrated gathering of extremely motivated, high IQ individuals that I've ever come across. This collective also has access to what at times feels like unlimited amounts of capital. Individuals here aren't afraid of risk and are highly networked. The result is fertile ground for ambitious projects to get started, including projects that simply wouldn't get funded in other regions, where the appetite for risk and access to capital is substantially inferior. Everyone is in Silicon Valley to build. Having said that, I do think the Bay Area's clear dominance will begin fade somewhat, as other regions are doing a better job on immigration, offer higher standards of living and are proactively incubating great talent.

Who have been role models you most admire?

He's not perfect, but I'm a fan of Chamath Palihapitiya. He's engaging and makes simple, yet hard to refute arguments. He's charismatic and has been successful on both the builder side as well as the capital allocation side. He made early, concentrated bets in outliers like AMZN, TSLA and Bitcoin. He's also pushing humanity forward with endeavors like Virgin Galactic.

What is the best advice you ever received about how to succeed in Silicon Valley?

Networking is about quality not quantity. Rather than networking extensively and casting the net wide, hoping to develop the biggest network, I have personally found it more fruitful to become true

friends with just a handful of folks within the larger tech network. Having a broad network of people that you barely know just isn't that useful. Their introductions are less potent, and they won't be there when you need them. The natural, organic, deep relationships are the ones that matter. These can't be networked or curated, they must come naturally. This network represents the people that will step up for you when needed, and you'll relish the chance to do business with them too. The rest is just a distraction.

What do you believe are the most important traits of a technology entrepreneur?

Resilience. Occasionally entrepreneurs get fortunate enough to enjoy a quick turn and exit fast, but in general, the startup game is a slog. You're attempting to defy the odds, break down barriers and succeed where others have no doubt tried and failed before. Competitors will push you to fail. Without grit, you're going to tap out before you reach the finish line.

What do you believe are the most important traits of an early stage technology investor?

Early stage investors need to develop deep conviction, even when others will be naturally cynical and doubtful about the prospects of early stage companies. Educated folk know that most of these startups will fail. The best ideas are often the most polarizing ones. When I first read about AirBnb, I thought it was a stupid idea, who'd want anyone sleeping in their living room? I'd have shared that view and clearly, I was very wrong. The best returns come from non-conforming deals and the spoils go to those who developed conviction early, and then held on, through thick and thin.

If you could tell a technology entrepreneur just one thing, what would it be?

Trust your gut. Stick to your guns. Even the best VCs, advisors and alike, will spend just a fraction of the time you dedicate to thinking about your business and sector. They're backing you to figure it out, so don't be swayed by the view or the advice of others. Definitely listen, absorb and learn from their advice where you can, but don't go against your gut if you fundamentally disagree. VCs are great at helping with the nuts & bolts of a business, but the vision part is really on you.

What do you know now that you wish you had known earlier on?

I learnt this lesson by working for someone that I should never have worked for. Sometimes the best option is to quit. I pride myself on grit and acknowledge that the top attribute I look for in an entrepreneur is resilience, per my prior answer. As such, this advice comes in direct conflict with that. Bear with me. I used to think I could make anything work if I tried hard enough, but ultimately, is it worth it? You have limited energy and must pick your battles. Sometimes the best option is to quit if the situation is lined up against you. Better to pivot, rather than fight endlessly, trying to make the impossible work. Put pride aside, a war of attrition is better not fought.

Would you please create a personal quote that captures important wisdom about participating in early stage investing?

"Follow the talent. Figure out who are the real builders, working on real-world problems and the rest will fall into place. Ignore the hype and the showmanship."

"Investing in early-stage startups can be like raising a child...however, the satisfaction of having achieved the impossible is unlike anything else."

Josef Holm

Founder, Venture Studio and Venture Capital Partner

Draper Goren Holm, Draper Venture Network

What did you want to be as a child?
Helicopter pilot.

What was your first career?
Technology Startup Entrepreneur.

What other careers have you had before your current one?
Crowdfunding Technology Provider.

Why do you choose to focus on early stage technology investing?
As a lifelong startup entrepreneur, I love the innovation-driven and fast-paced early-stage phase of companies and finding teams working on the next big thing.

When your investing is not going well, what do you do or say to yourself to keep motivated?
I try to analyze what went wrong and be grateful for the learning experience and do better next time.

What do you love most about what you do now?
What I love the most is playing a part in creating a better future by investing in socially impactful, game-changing technologies.

What dreams do you have for the next 10 years?
My dream is to achieve what we are working on right now at Draper Goren Holm, finding the smartest entrepreneurs and funding their brilliant blockchain projects to build the rails for more financial inclusion around the world. With a population of over 1.7 billion unbanked and billions more under-banked, the need for a system that works for everyone is glaringly apparent. In a future where access to the global financial system doesn't discriminate based on any marginality, instead, builds on math so that you don't have to trust banks or nation-state governments. By doing so, I believe everyone will be better off.

If you could change one thing about the world we live in, what would it be?
Money. The world deserves sound money; a fair, transparent, and easily accessible financial system that works for the masses instead of just the top 1%. Blockchain has shown a lot of potential in creating decentralized financial systems with Bitcoin leading the charge as a more secure, decentralized, immutable money hedge against inflation and a great store of value. All Fiat money in history has failed, and it is just a matter of time before monetary situations like that of Venezuela begin occurring in the USA and Europe.

What was your first technology investment and what happened?
We have a unique approach to vetting investment opportunities in the sense that we use our annual conference LA Blockchain Summit that is one of the leading blockchain conferences and takes place each October at the LA Convention Center (Except 2020 where the conference will be held virtually). Thousands of people attend the conference from around the world, and it is a perfect venue to meet startups and watch them interact with potential clients and investors throughout the two-day event. Amazingly, at every event, a few teams impressively stand out in the way they utilize the opportunities we offer them, like speaking on stage, doing media interviews, or pitching to investors.

The first company that we invested in impressed us so much on all the above that we decided to lead their seed round, and since then, the co-founder has become a venture partner, and we are working on groundbreaking new things together.

The bottom line is that our strategy revolves more around finding and investing in great teams that we are personally excited about. We focus on working with founders who we know will pivot with us along the way to reach our common goals, rather than just looking for outsized returns.

Can you share a story of one of your best investments, and why it went so well?

Our best investments have been those where we follow our gut instincts regarding the founding team rather than a polished presentation. In the early stage startup arena, everyone will have to pivot at some point and more important than the original business idea, or strategy is to invest in a team that recognizes the beast's nature and is willing to pivot with you and follow your guidance. Having been a startup entrepreneur for over 25 years myself, I have developed a pretty good sense of how to spot winning teams and weed out the sweet-talking time-wasters. It's like a BS detector that you sharpen by being in the trenches long enough yourself to know what it's like to hustle every day and if someone has what it takes.

Can you share a story of one of your worst investments and why it went so badly?

My worst investment decision resulted from going against my gut feeling. Always, always stay away from those, no matter what the numbers in some slick deck tell you. If it sounds too good to be true, it always is.

In your opinion, what is so special about Silicon Valley?

Nothing anymore. In the past, Silicon Valley has undoubtedly played a huge role in funding some of the biggest companies the world has ever seen but with ever-increasing connectivity and decentralization. It is no longer a must for startups to pack up where they are from to make a move to Silicon Valley to make it. I was born and raised in Austria, and I had to leave there 20+ years ago because of the non-existence of a startup community, angel networks, and VCs, which is no longer the case. Not only do US VCs actively scout Europe for deals, but there is also a growing number of European VCs and angel groups writing checks. In many ways, Silicon Valley has fallen victim to its success, and if COVID19 has taught us anything, we no longer live in a world where geographic location is a roadblock to global success.

Why do you think the innovation economy/Silicon Valley has a poor track record on diversity and inclusion? What needs to change?

It's not just Silicon Valley; the entire system is inherently racist. It's something that we are acutely aware of and have actively been trying to counter at Draper Goren Holm by paying extra attention to deals coming from minority groups and by making sure our events are inclusive to everyone. We've also sought advice and partnered with people like Rodney Sampson, who has not only become a Venture Partner at Draper Goren Holm but who actively advises us on how to walk the talk the right way and make an active impact. We're nowhere near where we want to be as an industry, but we also have to recognize that this process does take time and proactive change on many stubborn levels.

Some of this is conscious, and some are subconscious on the part of the investors. Either way, in 2020 and beyond it is no longer enough to be a non-racist or to be silent; one must be an anti-racist to counter the hate and pushback from groups that still believe in nonsense like white supremacy. Some of the issues will be solved by increased decentralization of organizations and making it less relevant to physically be in Silicon Valley or any specific geographic location to participate in the digital economy.

There is a lot more focus now by investors, employees and customers on companies that do good (and do no harm) as well as make money. How do you think about this topic in the context of your investing?

I think about social entrepreneurship as an integral part of any meaningful project. Naturally, I gravitate to startups who share my mindset of purpose and the common good over sheer profit. I vividly remember 2017 at the height of the ICO (Initial Coin Offering) craze when we were pitched countless projects that seemed to serve no purpose other than making as much money, which boiled down to nothing more than pump and dump schemes. We proudly stayed away from all of them, probably leaving millions on the table but feeling good about it. 99% of those ICO projects came crashing down spectacularly in 2018, and the take away is that success without purpose, whether it be the side of the VC or the entrepreneur, is entirely meaningless in the long run.

Who have been role models you most admire?

Being a high school dropout and struggling with the traditional education system, Richard Branson is one of my most significant role models. Throughout his school years, Richard had dyslexia, only to move on to become one of the most successful and inspiring entrepreneurs of our time. Other role models include Elon Musk, Tim Draper, and Steve Jobs, all of which have incredibly captivating stories that have profoundly shaped my way of thinking.

What is the best advice you ever received about how to succeed in Silicon Valley?

Never give up.

What do you believe are the most important traits of a technology entrepreneur?

The ability to recognize flaws in your original assumptions regarding your business model and readiness to pivot quickly (and multiple times if required) is the single most significant success factor besides never giving up.

What do you believe are the most important traits of an early stage technology investor?

Nothing is more important as an early-stage investor than knowing how to find the best teams building in an industry with incredible market potential. You don't have any past data on the company to rely on other than one's previous successes, failures, team dynamic, and being able to quantify the total size of the market they will penetrate.

If you could tell a technology entrepreneur just one thing, what would it be?

No guts, no glory.

What do you know now that you wish you had known earlier on?

Everything will work out just fine.

Would you please create a personal quote that captures important wisdom about participating in early stage investing?

"Investing in early-stage startups can be like raising a child. Many things can go wrong. There's great potential, but it will take countless course corrections and a never-give-up attitude to guide it to ultimate success. It's a terrible bet on the surface; however, the satisfaction of having achieved the impossible is unlike anything else."

> *"Go on the adventure with people you want to be with – and make sure you are wildly passionate about your mission."*

Linda Jenkinson

Serial Entrepreneur, Board Member & Scale Up Incubator Leader

Levelup, Air New Zealand, The Guild, Harbour Asset Management, Eclipx, Chair: Unicef

What did you want to be as a child?
To build bigger companies than my father.

What was your first career?
Strategy Consulting.

What other careers have you had before your current one?
Global Entrepreneur.

Why do you choose to focus on early stage technology investing?
As a women I felt it was really important to pay it forward and support all entrepreneurs who are going through the challenging "scale-up" phase.

When your investing is not going well, what do you do or say to yourself to keep motivated?
Lean in and provide strategic and "mental health" support. Opportunities come from tough times if you hang in and find them!

What do you love most about what you do now?
Being able to share what I have learned and be on the entrepreneurial journey with fabulous people doing great things!

What dreams do you have for the next 10 years?
To support 100 diverse entrepreneurs to build $100 million companies (50:50 female, male and culturally diverse).

If you could change one thing about the world we live in, what would it be?
To have all believe that protecting our environment is not an option and that if we all work together we can be economically well off in a beautiful, healthy & clean world.

Can you share a story of one of your best investments, and why it went so well?
My best investment in a company was where I ensured the following:

- A excellent CEO who was mentorable
- A very good board who was involved with the company and was diverse
- A great disruption opportunity
- I was in a role, Chair, where I could really positively influence and have a role where I could provide active guidance

Can you share a story of one of your worst investments and why it went so badly?
I invested in a company that took $ 6 million in investment and failed. As part of the process I took a board position. I noted that I was always in the minority on actions I thought the company was taking. The attitude I took was if the rest of the board and the founder did not agree with me, then it was ok for me to agree to disagree. As I looked back one of the reasons the company did not deliver on its promise – I realized that if I had taken more time to influence the rest of the board and the founder on the issues – the company may have had a better chance of success.

In your opinion, what is so special about Silicon Valley?
Silicon Valley lives and breathes growth companies. I remember when I first moved to the Valley I did not understand the acronyms people were using – everyone understands what it takes to grow companies and the support systems are there for growth.

Why do you think the innovation economy/ Silicon Valley has a poor track record on diversity and inclusion? What needs to change?
I don't 100% agree, I'm from New Zealand and being from anywhere is not a barrier in the Valley – where I do agree is that it is not easy being a women in the Valley. As a CEO I remember being asked by Venture Capitalist, was I fat or pregnant? Also most of the

more than $300 million I have raised has not been through traditional Silicon Valley channels because I did not fit that profile. I was outside the "boys club" and so was innovative in my approach. That is now starting to change and I am taking a lead in bringing about that change!

There is a lot more focus now by investors, employees and customers on companies that do good (and do no harm) as well as make money. How do you think about this topic in the context of your investing?

When I was 21, my mentor asked me what I wanted to do – and my response was "I wanted to make a positive difference in the world!" I say being an entrepreneur and doing good as linked. I have been involved in the social investing space for more than twenty years and do support the concept of sustainable development. I also see that the concept of for profit and non-profit are purely concepts – and that all enterprises are about succeeding to their mission and that in the "for-good" space it is even more critical to have excellent governance and a world-class team to successfully deliver to the mission. But every for-profit company I have started I have always integrated making a positive difference with achieving pure shareholder returns. To work so hard there has got to be multiple pay-offs, social and financial!

Who have been role models you most admire?

My father who was a serial entrepreneur, he started with nothing, and in a place with very little opportunity built multiple companies. He was very creative and taught me to define my own box.

What is the best advice you ever received about how to succeed in Silicon Valley?

I was not even aware of Silicon Valley until I got here – and discovered this incredible entrepreneurial mecca! The best advice I got was regarding networking. The Valley is 2 degrees of separation – and I learned that by forming a question and talking and exploring the question with people in the Valley within ten conversations you can find your answer. So you always take meetings and help people to pay it forward

What do you believe are the most important traits of a technology entrepreneur?

To be dauntless!

A open, learning mindset, resilience and determination.

To be an explorer and see the experience as a journey.

To stay in when times get tough.

What do you believe are the most important traits of an early stage technology investor?

To realize that you are not the CEO, and that your job is to be supportive of the entrepreneur on their journey.

If you could tell a technology entrepreneur just one thing, what would it be?

Don't give up – the difficulty is where the answers are…and HAVE FUN on the journey.

What do you know now that you wish you had known earlier on?

To have your own personal Board of Directors – a group of people who are just about supporting you.

Would you please create a personal quote that captures important wisdom about participating in early stage investing?

"We are going on an incredible adventure, we have a clear mission – but we have no idea what it will take and how we will get there – go on the adventure with people you want to be with – and make sure you are wildly passionate about your mission."

Do your homework always, never show up unprepared, you don't get many chances to impress and your knowledge about who you are talking to and why you are talking to them is critical to lead to a second conversation.

"In entrepreneurship, as in life, we have to give in order to get."

Regis Kelly

Director of Berkeley's Bakar BioEnginuity Hub Incubator (Opens July 2021)

Professor, UCSF; Professor, UC Berkeley

What did you want to be as a child?
Fighter pilot.

What was your first career?
Professor at UCSF.

What other careers have you had before your current one?
None.

Why do you choose to focus on early stage technology investing?
Innovation brought to reality through startup companies bringing benefits to society and local economic growth.

When your investing is not going well, what do you do or say to yourself to keep motivated?
We are not in it for the money.

What do you love most about what you do now?
Meeting highly talented and motivated people.

What dreams do you have for the next 10 years?
Reformatting public research universities so that they contribute to their public service mission by using their resources to help startups. Just as Academic Heath Centers promote local health by successfully combining academia with a business, running a hospital, so future Academic *Innovation* Centers would combine academia with incubators to promote society benefits and local economic growth.

If you could change one thing about the world we live in, what would it be?
Alter the aging process to that we stay healthy in mind and body till we die.

What was your first technology investment and what happened?
We started a tiny venture fund to provide financial support to startup life science companies that were too small to be of much interest to conventional venture funds managing hundreds of millions of dollars. Even although we did not have capital for follow on investments, the Fund is predicted to have about a threefold return, validating the concept.

Can you share a story of one of your best investments, and why it went so well?
One of our best investments was in synthetic biology. Because companies that used synthetic biology to make biofuels took such a beating, the appetite for investing in companies that wanted to persevere with synthetic biology was small. The leadership of the company we funded had learned their lessons from the failures of the biofuel industry and had pivoted to a new financial model by which they did not have to confront scale up problems and commodity pricing. They are doing VERY well now.

Can you share a story of one of your worst investments and why it went so badly?
Usually cells with problems die and their remains are quickly removed by cellular scavengers. Cancer cells have big problems of course but have become resistant to the processes that induce cell death. The company we supported was founded by well known scientists with excellent track records. They had a prototype drug that enhanced cell death in cancer cells. All their company had to do was to convert their compound by medicinal chemistry into a drug that could be used pharamaceutically. Unfortunately, despite heroic efforts they could not find any drug analog that retained the capacity to kill cancer cells. The company pivoted to another quite different target and became successful but this involved a down round in which our investment was essentially lost.

In your opinion, what is so special about Silicon Valley?
From my vantage point the Silicon Valley era had its origins in the flower power generation that dominated Northern California in the late sixties and early seventies. The ethos of those times tended

to reject capitalism and the accumulation of wealth, promoted community thinking and the importance of a balanced life and had a distaste for the hierarchical. These characteristics paved the way for community of innovators whose heretical thinking triumphed over the established, conventional companies then associated with Boston. In addition, California has a tradition of attracting immigrants, especially from Asian countries, and making them feel at home. The data linking entrepreneurship with immigration is so striking that it is tempting to believe that similar risk taking processes underlie starting a new company and deciding to leave one's home country.

Who have been role models you most admire?
William Rutter, the founder of Chiron, who continues to work in biotech, insightfully and successfully, into his 90s. Brook Byers, who has been willing to take the time to foster bridges between academia and the world of entrepreneurship, for the benefit of both.

What is the best advice you ever received about how to succeed in Silicon Valley?
Get to know a lot of people. You never know when it will be useful.

What do you believe are the most important traits of a technology entrepreneur?
I like to see a burning passion in the entrepreneur for their company to succeed, an enthusiasm for working as a team member and a willingness to listen carefully to advice.

What do you believe are the most important traits of an early stage technology investor?
Since in early stage investment the risk is often mainly technical, not managerial or financial, it is great if the early stage investor has the capacity, through education or experience to evaluate the technical risk. Assuming the innovator is giving an unvarnished description of the science and technology behind the innovation is risky.

If you could tell a technology entrepreneur just one thing, what would it be?
Make sure that you have a knowledgeable CEO.

What do you know now that you wish you had known earlier on?
As a standard cookie-cutter academic I had little appreciation of how partnership with, and respect for, the private sector is crucial if we are to turn our discoveries into products and services of value to society. And to appreciate that conversion of our discoveries in to something useful was essential if we hoped that our basic research would be financed by tax-payer dollars.

Would you please create a personal quote that captures important wisdom about participating in early stage investing?
"In entrepreneurship, as in life, we have to give in order to get."

Why do you choose to focus on early stage technology?

I believe entrepreneurship can break the cycle of poverty and transform lives and communities.

"It is better to be lucky than to be smart. But even to be lucky you have to be in the right place at the right time with the right people/team to realize that luck."

Greg Kidd

Investor and Co-Founder

Hard Yaka and GlobaliD

What did you want to be as a child?
A leader.

What was your first career?
Outdoor Trip Leader.

What other careers have you had before your current one?
Management Consultant, Senior Analyst Federal Reserve, Risk and Regulatory Advisor/Consultant.

Why do you choose to focus on early stage technology investing?
This is where efforts occurred for things I wanted to see in the world in my short lifetime.

When your investing is not going well, what do you do or say to yourself to keep motivated?
Find other investments and ride my bike. Also invest in my own efforts and build new products to seek product market fit.

What do you love most about what you do now?
Constantly learning and get the chance to build things with my peers that make a difference in the world.

What dreams do you have for the next 10 years?
Create self-sovereign identity where people are empowered rather than trapped in silos run by governments or corporations. Financial inclusion for all.

If you could change one thing about the world we live in, what would it be?
Better education access to create true equal opportunities and appreciation for democratic civil society.

What was your first technology investment and what happened?
First tech investment that was not my own company is Twitter. It has become the most influential political publishing platform in the world. Allowed my financial independence.

Can you share a story of one of your best investments, and why it went so well?
Negotiated 10 years warrants on 1% of the float of all XRP (the worlds 3rd ranked crypto after Bitcoin and Ethereum). Luckily or correctly foresaw the possibility that crypto would be a legitimate segment. Rode the wave and enabled funding of my other visions for necessary infrastructure changes (in identity) to make the promise of a digital financial world safe and compliant.

Can you share a story of one of your worst investments and why it went so badly?
3taps scraped public data (Craigslist) and repurposed as API for third parties like PadMapper. The offering worked but was a magnet for litigation that nearly bankrupted me with legal fees. I should have stopped the business and fought the lawsuit on offense rather than defense.

In your opinion, what is so special about Silicon Valley?
Attracts the best talent and puts us all in the same fishbowl so that we can easily cross-fertilize. Failure in Silicon Valley = Experience and is appreciated as such.

Who have been role models you most admire?
Jack Dorsey for showing me the difference between an idea (in the head) and an insight (in the stomach and the heart). True transformation requires the later not the former.

What is the best advice you ever received about how to succeed in Silicon Valley?
Forest Gump: life is a box of chocolates – you never know what you are going to get. The ability to suspend disbelief. That you can wager on things that don't statistically make sense but that can pay off astronomically even though they don't.

What do you believe are the most important traits of a technology entrepreneur?

Having a vision of what the world should be like rather than could be like, and having the resilience in a Gandhi like way to be the change you want to see in the world.

What do you believe are the most important traits of an early stage technology investor?

Willingness to evaluate people rather than just the idea…and willingness to accept that 7 out of 10 things back will fail outright within 2 years of funding.

If you could tell a technology entrepreneur just one thing, what would it be?

Make sure you have an insight and not just an idea.

What do you know now that you wish you had known earlier on?

Don't take all the success or failure personally. You matter but you matter in the role you are playing. You are not your product/service.

Would you please create a personal quote that captures important wisdom about participating in early stage investing?

"It is better to be lucky than to be smart. But even to be lucky you have to be in the right place at the right time with the right people/team to realize that luck."

> *"The best early stage investments are made when you have a strong point of view about where the world is headed in the next ten years, how it will start to get there in the next three, and who is working on it now."*

Ian Lee

Venture Capital Partner
IDEO CoLab Ventures

What did you want to be as a child?
An artist. I was a studio fine art major and studied Renaissance art and architecture in college. I eventually switched to human-computer interaction and economics, which led me into the fields of technology, business, and finance.

What was your first career?
I was a management consultant for six years. I spent a lot of time in the mid-west working with large aerospace and automotive companies on growth, operational turnaround, and manufacturing optimization projects. I worked on the V-22 Osprey for a year.

What other careers have you had before your current one?
Advertising, marketing, investment banking, management consulting, corporate innovation, and corporate venturing.

Why do you choose to focus on early stage technology investing?
I've always loved imagining what the world will look like and figuring out how we'll get there. This led me to early stage technology investing.

When your investing is not going well, what do you do or say to yourself to keep motivated?
If you're not failing or making mistakes, you're not growing. What's important is that you learn and get better from them.

What do you love most about what you do now?
The privilege to meet and work with incredibly talented entrepreneurs on some of the most important, difficult, and valuable problems in the world.

What dreams do you have for the next 10 years?
People think of startups, venture capital, corporate venture capital, and corporations as these things that exist in isolation, separate from one another, when in fact, they're different organizations designed to do the same thing at different speeds and scales: to improve the world we live in and serve people's needs. That means these organizations are on a collision course, and in the next ten years, we will see more partnerships between startups, venture capital firms, and corporations, as well as the emergence of new, integrated models for venture building that people today think are impossible or impractical. I believe the future of venture is collaborative—and that collaborative models for venture building will become dominant over traditional models of venture.

If you could change one thing about the world we live in, what would it be?
Over the last several decades, wealth has disproportionately accrued to the creators of (and investors in) technology platforms and away from labor, particularly as technology has automated jobs and made fewer people more productive. I don't believe this is sustainable long term for people or society. I believe that everyone will become co-owners in the technologies, products, and services they use—not only after a company goes public but at the earliest stages of their development. Decentralized technologies like crypto will enable this—this is what I've been investing in and working on for the last half decade.

In your opinion, what is so special about Silicon Valley?
There are so many reasons why a startup can fail, and most come down to a lack of access to things like talent, customers, distribution, capital, and knowledge. This is why Silicon Valley is so special: it has such a high concentration of all of the things that startups need that it gives entrepreneurs here better odds at succeeding. Higher rates of success bring more talent, customers, capital, and experts to Silicon Valley, further improving future outcomes and odds. It's a virtuous cycle that has been accumulating for decades, and it is not easy to recreate.

There is a lot more focus now by investors, employees and customers on companies that do good (and do no harm) as well as make money. How do you think about this topic in the context of your investing?

For years, "doing good" was viewed as something non-core to a business and its strategy—it was primarily for philanthropic or marketing purposes. However, as consumers and employees alike have become more conscious of the impact their choices make on things like the environment, community, people, and even their own self identities, they've come to expect the same from the companies they give their money, time, and talents to. It's no longer sufficient for a company to make a good product and deliver profits to shareholders—the company needs to hold itself to a higher standard and realize that it is accountable to a broader set of stakeholders, not just its shareholders, to succeed. In this way, "doing good" is not just good PR, it's now critical to a business' strategy to grow and survive. When investing in startups, I work with entrepreneurs to understand the contexts in which they live and the players they must both compete and collaborate with. Companies don't just operate in markets. Companies today operate in complex, interconnected ecosystems.

What is the best advice you ever received about how to succeed in Silicon Valley?

"The slope of the line is more important than the y-intercept."

Would you please create a personal quote that captures important wisdom about participating in early stage investing?

"The best early stage investments are made when you have a strong point of view about where the world is headed in the next ten years, how it will start to get there in the next three, and who is working on it now."

> *"Nobody is good at picking winners early on. If you are lucky a winner will emerge from your portfolio and if you are really lucky you will get the chance to double down."*

Duncan Logan

Incubator Founder, Angel Investor

Nex Cubed

What did you want to be as a child?
Farmer.

What was your first career?
Derivatives Trader.

What other careers have you had before your current one?
Entrepreneur and Sales leader.

Why do you choose to focus on early stage technology investing?
I love people who are brave enough to become entrepreneurs, I see the possibility in startups.

When your investing is not going well, what do you do or say to yourself to keep motivated?
The highs are never as good as they seem and the lows are never as bad.

What do you love most about what you do now?
I love being on the edge of the future. I love trying to understand new concepts, ideas, innovations.

What dreams do you have for the next 10 years?
I have another startup in me, I would love to take all the learnings from my life so far and build another business. I also want to build up a portfolio of investments perhaps 100 different companies where I can deploy some capital and then help out using my network to make them successful.

If you could change one thing about the world we live in, what would it be?
The impact of climate change is going to become more and more obvious. The hardships climate change will impose on billions of people are hard to imagine today but will become reality.

What was your first technology investment and what happened?
Like so many investments, it is still limping along 10yrs on. It has neither thrived nor died but instead tied up a great entrepreneur who feels compelled to make it work. I know the entrepreneur personally and have watched the immense impact of this venture on his health, marriage and children. It reminds me of how choosing to be an entrepreneur can affect everyone around you. It is not for the faint hearted.

Can you share a story of one of your best investments, and why it went so well?
I invested in a great entrepreneur who struggled like mad before things began to go his way. I helped him through the tough times and then the company was approached to be purchased.

All the minority investors tried to stop the sale except me. The investors got the cash from the deal and the entrepreneur was forced to take stock in the acquiring company that I agreed to take as well. The stock multiplied 15x in the earnout period making it a great investment.

Can you share a story of one of your worst investments and why it went so badly?
Nope…too painful.

In your opinion, what is so special about Silicon Valley?
The level of curiosity is obscene. It does not matter the industry, from funerals to farming, crypto to camping there are entrepreneurs trying to work out how to innovate to build a business. The more obscure the more interesting. I love it.

Who have been role models you most admire?
Elon Musk for his enormous goals and simple thinking, Mark Andreessen for his straight talk and ability to explain things, David Goggins for his drive.

What is the best advice you ever received about how to succeed in Silicon Valley?
Don't build a new company, build a new category.

What do you believe are the most important traits of a technology entrepreneur?

Curiosity and tenacity.

What do you believe are the most important traits of an early stage technology investor?

The ability to ask good questions, and patience.

If you could tell a technology entrepreneur just one thing, what would it be?

It takes a team to build a company, as the entrepreneur your greatest responsibility is building the best possible team and then ensuring they work effectively to a clear vision. Then step out of the way.

What do you know now that you wish you had known earlier on?

I wish I had learnt to code.

Would you please create a personal quote that captures important wisdom about participating in early stage investing?

"Nobody is good at picking winners early on. If you are lucky a winner will emerge from your portfolio and if you are really lucky you will get the chance to double down."

"Investing is as much an art as it is a science. Qualitative is as important as quantitative."

Chin-Chai Low

Angel Investor

Berkeley Angel Network, Keiretsu Forum

What did you want to be as a child?
Engineer.

What was your first career?
Software Engineering in Communications Industry.

What other careers have you had before your current one?
Real Estate Lending and Real Estate Investing.

Why do you choose to focus on early stage technology investing?
Early stage investing provides early access to highly promising companies, provides the highest return and I enjoy engaging with entrepreneurs directly.

When your investing is not going well, what do you do or say to yourself to keep motivated?
Enjoy the process, not just the end result. Learn quickly and adapt. Failures are part of the game and successes takes time.

What do you love most about what you do now?
Learning about new innovations, networking with other investors and interacting with entrepreneurs.

What dreams do you have for the next 10 years?
My goal is to show I can be successful in early stage investing and continue to do this for a long time. Although success is ultimately measured by financial returns, it is also just as important to me to have made a difference in the companies I invest in and to have these companies have positive impact on the lives of people. In order to be able to do this effectively, it is

essential that I can build a network of companies and investors that will allow me to add value and increase the chances of success of companies I invest in.

If you could change one thing about the world we live in, what would it be?
Our world will be a much better place if our government adopt policies that are based on individual responsibility and accountability practiced with a combination of wisdom, honesty, kindness and compassion.

What was your first technology investment and what happened?
I invested in a company that was a pre-cursor to Software-As-A-Service (SAAS) back in dotcom days. Its approach was to host existing software on centralized servers that are accessed using remote desktop software. It did not gain a lot of traction and was merged with another company that took the assets and I never heard from them again. You could say I was early with the idea of SAAS as Salesforce was started at about the same time. We had the right idea but the wrong approach.

Can you share a story of one of your best investments, and why it went so well?
This story is still being written as most of my investments were made in the last 2-3 years and none has exited. However, the investment I am most excited about is a company called Blok Party that makes electronic board game platform PlayTable. The company was founded and led by a very young entrepreneur with incredible creativity. The company has secured licensing agreements to develop electronic versions of many popular physical board games and bringing new innovations to these games that are made possible by PlayTable. The company has shipped limited quantities so far but will be ramping up the volume this year, so this is going to be an exciting year. I believe PlayTable will be the next new gaming platform.

Can you share a story of one of your worst investments and why it went so badly?
This was in a company called MonkeyBin. This is another dotcom investment. The company was attempting to improve efficiency of bartering to via multi-party exchanges using matching algorithms. In hindsight, it was a silly idea because it was not solving a problem anybody had. The problem of bartering was already solved by the invention of

currencies. Safe to say I really did not know what I was doing in those days and thankfully I made very few investments.

In your opinion, what is so special about Silicon Valley?

Silicon Valley is a unique ecosystem of people, money, technology, companies, and top universities in the world. The people here are the entrepreneurs, the technical talent, and investors. It draws in people from other parts of the country and the world like no other places, bringing their talents and ideas, and nurtured by existing companies, angel investors and venture capital firms. The concentration of all these elements makes things start faster, move faster, fail faster and succeed faster.

Why do you think the innovation economy/ Silicon Valley has a poor track record on diversity and inclusion? What needs to change?

Making changes to diversity is like making changes to anything else. There is always inertia and the need to be out of our comfort zone. Those already there does not have natural inclination to change as there are usually no obvious benefits to them personally. So incentives matter.

Personally, I have worked with many women engineers and have hired many. The engineering teams I have managed often has the largest proportion of women than any other engineering teams I know. And this was about 20-30 years ago when gender diversity in engineering was not in most people's minds. I did not hire them because they were women and diversity was not my goal. I just hired people that I thought were the most qualified and would work well in my team. My incentive was to have the best team members. I only realized all of these years later when the subject of diversity became a hot topic.

So to answer this question I think one has to also define the goal and purpose of diversity. If you ask this question to any organization, the usual answer is there should be more because it is better but cannot say what the end goal should be or why that is the right end goal. Often the answer given is a one size fits all answer: it should reflect the population. For example does gender diversity mean there should be 50% women in every kind of job? Why is this the right goal for every kind of job? We often hear there aren't enough women engineers but is 50% always the right number? Should 50% of plumbers also be women? Should 50% of nurses and teachers be men?

If not, why not? If one poses the same question using other common characteristics for diversity such as race and sexual orientation, the problem becomes intractable and we are likely to fail. So rather than focusing purely on diversity for diversity sake, we should focus on the tangible return of diversity and put in the right incentives to achieve it.

There is a lot more focus now by investors, employees and customers on companies that do good (and do no harm) as well as make money. How do you think about this topic in the context of your investing?

I absolutely believe in this philosophy. I believe in Karma – what one does will eventually affect our own wellbeing. So doing good while making money is just natural. I also believe that it leads to better results as when you do good you are also solving a need. That does not mean that all good things are good investments. They must also provide an adequate financial return for the risk. For doing things that don't meet my criteria of financial return, I do that through charitable work. It's important for me to separate investing from charity.

Who have been role models you most admire?

The Buddha is my most important role model. He teaches us that all beings have a Buddha nature and we are all the same. What differentiates us are our afflictions that arises out of our ignorance which causes us to suffer. Removing our afflictions removes our suffering and make us all the same. The path to removing our afflictions and suffering is through wisdom, kindness and compassion.

What is the best advice you ever received about how to succeed in Silicon Valley?

Focus on helping others to succeed and success will follow.

What do you believe are the most important traits of a technology entrepreneur?

1. Persistence, endurance, being focused.
2. Ability to communicate clearly.
3. Ability to learn fast and fail fast.
4. Ability to attract talent.

What do you believe are the most important traits of an early stage technology investor?

1. Have clear criteria for investments.
2. Have a consistent process.
3. Have patience.

4. Bring a strong network of investors and companies to add value.

If you could tell a technology entrepreneur just one thing, what would it be?

The best type of business is one in which its moat gets stronger as it gets bigger.

What do you know now that you wish you had known earlier on?

True joy comes from helping others.

Would you please create a personal quote that captures important wisdom about participating in early stage investing?

"Investing is as much an art as it is a science. Qualitative is as important as quantitative."

"Trust your instinct and forget about it after writing the check."

Jennifer Lui

Angel Investor and Venture Capital Partner
Surfbird Investments, Newport Private Group

What did you want to be as a child?
Teacher.

What was your first career?
Business Analyst.

What other careers have you had before your current one?
International Business.

Why do you choose to focus on early stage technology investing?
Challenging and exciting.

When your investing is not going well, what do you do or say to yourself to keep motivated?
There will always be another one.

What do you love most about what you do now?
Seeing people from all backgrounds achieve their dreams.

What dreams do you have for the next 10 years?
Eco friendly and sustainable environment.

If you could change one thing about the world we live in, what would it be?
Reduce wastes, particularly plastic.

What was your first technology investment and what happened?
Acted as an angel investor for a big data company. The company received series B funding and is doing well.

In your opinion, what is so special about Silicon Valley?
Failures are welcomed.

What is the best advice you ever received about how to succeed in Silicon Valley?
Keep trying.

Would you please create a personal quote that captures important wisdom about participating in early stage investing?
"Trust your instinct and forget about it after writing the check."

"It doesn't take a miracle to be successful. You need a quality product or service, a vetted plan and a supportive network of people who can provide honest, advice, guidance and encouragement."

Sharon Miller

CEO Renaissance Entrepreneurship Center

What did you want to be as a child?
Archaeologist.

What was your first career?
Executive Vice-President of the non-profit organization, American Jewish World Service which provided non-sectarian humanitarian aid, technical assistance and disaster relief to grassroots organizations in the developing world.

What other careers have you had before your current one?
I co-owned a wholesale travel business.

Why do you choose to focus on early stage technology?
I believe entrepreneurship can break the cycle of poverty and transform lives and communities.

When your work is not going well, what do you do or say to yourself to keep motivated?
Although not everyone who hopes to start businesses succeeds, they are developing life-enhancing skills, confidence and networks.

What do you love most about what you do now?
Entrepreneurs are the most resilient, courageous people on earth. They are visionary and creative and inspire me with their ingenuity, stubbornness and grit.

What dreams do you have for the next 10 years?
I would like to see a world where economic opportunity, resources and capital are broadly available to individuals regardless of their race, gender, immigration status, or income level. Entrepreneurship is a critical means for women and men from under resourced communities and populations to achieve economic mobility, build assets and create better futures for themselves and their families. Unfortunately, there are massive numbers of people with great ideas, integrity, passion, a commitment to hard work, who do not have access to the tools and resources they need to succeed.

If you could change one thing about the world we live in, what would it be?
I would like to end poverty – women, men and children around the world should all live with dignity. That starts with the ability to fully support yourself and your family.

Can you share a story of one of your best investments, and why it went so well?
My organization, Renaissance Entrepreneurship Center has had the privilege of working with Yvonne Hines, a lifelong resident of the Bayview Hunters Point community in San Francisco. She is the proud owner of Yvonne's Southern Sweets. Yvonne, who previously worked as a Meter Maid, was pregnant with her first child and wanted to start a business to help her pay for child care. She turned to what she knew how to do best. She used her grandmother's recipes and began baking mouthwatering pralines and butter cookies and started to sell them out of her car to members of her church. Yvonne and her southern desserts were so well received, she was ready to follow her big dream and open her own bakery. Yvonne came to Renaissance and we worked with her - every step of the way - from reviewing the feasibility of her idea to providing training in operations, finance and marketing, and assistance in developing her website and using technology. We were also able to assist Yvonne in securing the capital she needed to help her develop marketing materials, upgrade the building's façade and purchase equipment. Fifteen years later, Yvonne's Southern Sweets on Third Street in Bayview Hunters Point, San Francisco is a neighborhood institution. Yvonne is a leader in her community, serves as a role model to other women seeking to start businesses and Yvonne's Butter Cookies and Pralines are sold at the Warriors

Stadium among many other locations. Her daughter, who served as her "Assistant Manager" as soon as she could talk, is now a high school senior, top of her class and hopes to attend Stanford or Howard University in the fall 2021.

There is a lot more focus now by investors, employees and customers on companies that do good (and do no harm) as well as make money. How do you think about this topic in the context of your investing?

Creating equity, promoting sustainable economic opportunity, working in community and supporting self-determination and dignity are organizational values that inform Renaissance's programs and services. Our clients overwhelmingly share these values and often refer to them as the inherent reasons they want to own and operate their own businesses. Our clients create businesses that are frequently focused on solving local problems, and bringing new goods and services to communities and populations that, due to their demographics, were previously not serviced by others. Further, often as a result of their own negative experiences, our clients seek to develop businesses where all employees, customers, and vendors are treated with respect and dignity. From giving people their first jobs, to hiring others who are returning to the workforce, to mentoring and being role models, these factors differentiates our entrepreneurs in a very positive way. As one client states "I do this to show the kids on the hill that if I can have my own business, so can they." As they support and engage directly with their communities, leading with their values, our small businesses earn respect, customer loyalty and a commitment to help them to succeed.

Who have been role models you most admire?

Alison Davis. She is brilliant, gracious, and an expert problem solver who generously devotes her time and resources into creating solutions to address economic inequality.

Would you please create a personal quote that captures important wisdom about participating in early stage investing?

"It doesn't take a miracle to be successful. You need a quality product or service, a vetted plan and a supportive network of people who can provide honest, advice, guidance and encouragement."

"A great investor backs companies that build products that count: products that transform business and life at scale."

SC Moatti

Venture Capital Partner

Mighty Capital

What did you want to be as a child?
Engineer.

What was your first career?
Product Manager.

Why do you choose to focus on early stage technology investing?
I love innovation and anything that can make us better people.

When your investing is not going well, what do you do or say to yourself to keep motivated?
You are ready.

What do you love most about what you do now?
Spending the majority of my time in my "Zone of Genius" – look this up if you don't know about this awesome concept!

What dreams do you have for the next 10 years?
Redefine the role of venture capital (VC) in shaping economic growth. I believe that VC is being disrupted by two major shifts. First, there has been a consolidation of incumbents into mega-funds. Second, we've seen a blurring of public and private equity pioneered by unicorn fundraisers and crypto-currencies. This puts VCs in a classic innovator's dilemma: if I keep getting bigger in order to survive, my returns will get diluted which is ultimately lethal. The solution is to change the rules of the game – which is what I set out to do with Mighty Capital.

Can you share a story of one of your best investments, and why it went so well?
We invested in an analytics company with an amazing management team, a fantastic product and great co-investors. Every year, the company was able to deliver on its key targets and doubled its valuation.

The founding CEO invited us to participate in the deal because of the value of our global product network. What compelled us to invest in addition to the fantastic traction, is that the company is really changing the way business is done by providing transparency and simplicity to our entire economy and enabling innovation in both large and small organizations. It eventually became a unicorn and went public.

Can you share a story of one of your worst investments and why it went so badly?
We invested in a gaming company during a severe recession. Within a month of investing, the CEO quit for a cushy job at a large company. It was a complete surprise to all. In an interview, they bragged about all they were going to do at this new company that they couldn't do before. It was so unethical! To top this of, the board decided to keep the company going without addressing the elephant in the room, so executives and team members started to question why they were here themselves. We were able to pull the plug and get our money back.

Who have been role models you most admire?
Jeff Bezos, Christine Lagarde and Susan Mason.

What is the best advice you ever received about how to succeed in Silicon Valley?
You are ready.

What do you believe are the most important traits of a technology entrepreneur?
Great salesmanship.

What do you believe are the most important traits of an early stage technology investor?
Investor instinct, attention to detail, and relationship building.

If you could tell a technology entrepreneur just one thing, what would it be?
Your work matters: 95% of jobs are created by companies 5 years old or younger.

What do you know now that you wish you had known earlier on?
You are ready.

Would you please create a personal quote that captures important wisdom about participating in early stage investing?
"A great investor backs companies that build products that count: products that transform business and life at scale."

"Early stage venture capital is a long game. You have to have a lot of patience and the ability to ride the ups and downs"

Shayna Modarresi

Venture Capital Partner and Angel Investor

Lodestar Ventures, Correlation Ventures and Band of Angels

What did you want to be as a child?
Medical Doctor.

What was your first career?
Investment Banker (M&A and private placements).

What other careers have you had before your current one?
Entrepreneur. I also spent a few years raising institutional capital for venture funds as a placement agent.

Why do you choose to focus on early stage technology investing?
I love working with entrepreneurs when they are just getting started. To me, it's much more exciting at the earliest stages when very few others believe in the company. I also think there is a special bond that you build with an entrepreneur being one of the early believers and a trust that's built which is hard to replicate at the later stages. You experience all the ups and downs of a business just getting off the ground. You become their therapist, coach and cheerleader.

When your investing is not going well, what do you do or say to yourself to keep motivated?
This is a great learning experience. I will get some investments wrong, but venture is about the 4 you got very right and not the others that didn't go as well as expected.

What do you love most about what you do now?
I love connecting and working with entrepreneurs; seeing all the ideas. I also love helping companies grow into something big is very rewarding, both personally and financially. There is a sense of ongoing accomplishment.

What dreams do you have for the next 10 years?
Continuing to invest in startups, see more successes and grow my venture fund.

If you could change one thing about the world we live in, what would it be?
Global Warming. We need to secure a sustainable path for future generations.

Can you share a story of one of your best investments, and why it went so well?
Some businesses are overnight successes but few of them really move immediately up and to the right. One of my best investments out of Lodestar Ventures is a company that built an artificial intelligence platform that helps organizations raise capital by increasing their efficiency with actionable data analytics. They use intelligence to predict which donors will make the next big gift, assisting fundraisers in focusing on the best donors at the right times throughout the year to maximize the revenue they generate. The company exited in 19 months after our initial investment giving us a 4.6x return and an IRR of 154%. I met the founders through another investor who is on the east coast. It's worth noting that the company is headquartered in Boston and if it was not for this particular investor sharing the deal with me I may not have seen it. This is why the strength and quality of your network matters. Why it went so well: They are an extremely passionate founding team with previous experience in the industry they are tackling. The founders also complemented each other well. They also had an early movers advantage.

Can you share a story of one of your worst investments and why it went so badly?
My worst investment was in a very promising on-demand start up in the consumer space. It was a hit with customers and took the world by storm. The founding team had experience navigating the market and created the company out of their own pains. It had a strong founding team and, eventually, a "who's who" of Silicon Valley investors. The company raised north of $80M before being forced to shut down. Why it went badly: The founders believed

that their success was assured given the need for the service (it was a great service!). I believe the founders got too wrapped up in a particular idea and its implementation details. This obfuscated the product/market fit and prevented necessary pivots. They had a difficult time establishing a solid business model from the beginning and the need for rigorous planning and foresight. As time progressed, it also became evident that the CEO struggled to take input from the rest of the executive team. Lastly, they ignored what their investors and advisors were telling them which proved to be fatal. They ran out of runway with no investor buy-in on the future of the business.

What do you believe are the most important traits of a technology entrepreneur?

Perseverance and passion; giving their all to the startup. Almost like an obsession. Entrepreneurship is a game of attrition. It's about having the determination and discipline to see it through.

What do you believe are the most important traits of an early stage technology investor?

1. Strong gut-based decision-making: Investing at an early stage there is often little to no data that points to the success of the business. As an investor, you're really making an investment in the founding team and need to have the ability to identify talent.

2. Network abilities: The ability to source proprietary deal flow through your network. Venture capital is a people business. Having a strong network that trusts you is crucial to your success.

3. Patience: Investing at an early stage requires a lot of patience. The average time to exit is around 8 years.

At the end of the day, the ability to roll up your sleeves and become a therapist, coach, and cheerleader to the companies you invest in.

If you could tell a technology entrepreneur just one thing, what would it be?

Be passionate about your vision. Early stage venture is a lot about the entrepreneur - who that person is and what she or he is capable of.

Would you please create a personal quote that captures important wisdom about participating in early stage investing?

"Early stage venture capital is a long game. You have to have a lot of patience and the ability to ride the ups and downs"

"The long game is the most satisfying game"

Syd Mofya

Venture Capital – Global Alliances

Draper Venture Network

What did you want to be as a child?
Doctor.

What was your first career?
Financial Audit.

What other careers have you had before your current one?
Chemical Engineer in oil industry, Program Manager in public health & energy industries.

Why do you choose to focus on early stage technology investing?
It is exhilarating to work with the innovators who are shaping the future of our world.

When your investing is not going well, what do you do or say to yourself to keep motivated?
Note: the Draper Venture Network entity does not invest in companies. I speak in the capacity of my personal investments. I gain perspective: Important as investing is, there are more important things in life. I also try to capture any key lessons for the next time.

What do you love most about what you do now?
Being at a nexus of activity and connectivity - working with innovators and investors all over the globe is a constant source of joy and hope!

What dreams do you have for the next 10 years?
My wife and I just had a daughter (April 2020) in the midst of the global covid pandemic and an important global conversation about justice and equality. In 10 year's time, she will be 10, and my older daughter will be 23. I dream for them to enjoy more freedom to pursue what they would dream to pursue, wherever in the world they would like to do it – without experiencing prejudice. I dream that the liberty they and we now have will become true freedom. Liberty is enshrined in the words of the U.S constitution, and freedom is liberty in action. Technology plays a big part in whether or not this kind of world becomes a reality.

If you could change one thing about the world we live in, what would it be?
I would make it easier for us all to focus more on the things that matter most. Money makes the world go round, and it is a really great measure of value, and an enabler of exchange. It also demands more and more time, which is the real limited resource. If technology (e.g. AI, blockchain, etc) can help us to get to a place where we spend more time with families and friends, and to have a healthier relationship with our own bodies and souls - all this work would have been worthwhile.

What was your first technology investment and what happened?
My first technology investment was when I started a computer business in Zambia that took advantage of a large difference in import tax rates between computer parts and fully assembled computers. We imported the parts, assembled the computers in Zambia, and sold them to businesses, and occasionally individuals. I invested any extra cash I had as a student working several jobs in this business, bootstrapping it to profitability. The business itself never grew big enough to justify a healthy exit, but the business lessons I learnt from that experience are valuable to me to this day. It's where I first really learnt the importance of having the right teams, who connect at a level more fundamental than just the business. Alignment of values is critical for any endeavor I am a part of, and has become an important lens for me when evaluating opportunities. So, in that sense, I would say it was a successful investment!

In your opinion, what is so special about Silicon Valley?
I moved to Silicon Valley in 2014 from Tanzania – where I had started on the early journey of creating a venture fund to invest in African early-stage tech entrepreneurs. Within three weeks of being in Silicon Valley, I had met more people doing tech investing in Africa here than I had in four years in Tanzania. The network of talent and ideas is denser than anywhere else I have lived, and it is also amazing how willing people are to introduce others who could be of value to each other. It creates this vortex of increasingly rich networks – a form of the "Matthew Effect" where the more you have, the more you get.

Why do you think the innovation economy/ Silicon Valley has a poor track record on diversity and inclusion? What needs to change?

I think ultimately it comes down to a structural imbalance that has never been addressed with the kind of energy it needs to resolve it. Previously the "incumbents" would say it's a pipeline issue, but it has become clear that that is not where the root problem lies. The root problem, in my view, lies in the fact that over many years, Silicon Valley has been "successful" without embracing diversity and inclusion.

Most people I meet in Silicon Valley are not racist, but the system is racist by default, because it, by default, excludes people who are not already in the "network" and they have to work twice as hard to get a seat at the table everything being equal. This is another manifestation of the "Matthew Effect" – those who already have, simply get more. So how do you counter it? It takes a whole lot of force to counter a natural phenomenon. Frankly, it's not enough of a problem for enough people. I see shoots of hope though, and as we get more women and people of color in influential positions, perhaps we may get to a tipping point where there are enough people for whom change is a must. That said, there is a lot of work to do.

Who have been role models you most admire?

My late Dad – Visionary, wise, a model of resilience and enterprise.

Nelson Mandela – Courageous in adversity, Magnanimous in dealing with his enemies.

Quincy Jones – A pioneer, with such longevity in the music industry.

What is the best advice you ever received about how to succeed in Silicon Valley?

Bring value. Lead with value. Give first (all the different variations).

What do you believe are the most important traits of a technology entrepreneur?

Deep passion for what they do. A "secret" that gives them an advantage over others trying to do the same thing. Ability to build a team.

What do you believe are the most important traits of an early stage technology investor?

Access to the best deals

Being someone entrepreneurs want to team up with (which is another way of saying #1)

What do you know now that you wish you had known earlier on?

Time is our most valuable resource. In my 20s and perhaps 30s, I didn't think it was a big deal, but now, in my 40s I see that time truly is of the essence.

When my Dad died relatively young, it dawned on me that our time on earth is not infinite. I'm trying to make it count.

Would you please create a personal quote that captures important wisdom about participating in early stage investing?

"The long game is the most satisfying game"

I would give the entrepreneur the same advice I once received: manage your attitude and never give up, even in the darkest of times. Never give up and learn from your mistakes.

> *"Early stage investing is all about relationships and diversification. Do your homework in learning about the team."*

Maryanne Morrow

CEO, Founder and Angel Investor

9th Gear Technologies, Inc. and Keiretsu Forum

What did you want to be as a child?
International businesswoman.

What was your first career?
Registered Representative for AXA selling financial products and investments to individuals.

What other careers have you had before your current one?
Institutional Finance—1. Business development for attribution software that we sold to S&Ps; 2. Designing fee-based asset management platforms for banks, broker dealers and insurance companies that we sold to BNP Asset Management; 3. Running the custom content group for Dow Jones and 4. Multimedia account executive concentrating on financial firms for The Wall Street Journal.

Why do you choose to focus on early stage technology investing?
It's fascinating to see the journey from an idea into a commercialized business.

When your investing is not going well, what do you do or say to yourself to keep motivated?
Diversification keeps me sane. Picking up the phone to help startups navigate the rough patches saves investments. Sometimes a great idea is before its time; just chalk it up and move on.

What do you love most about what you do now?
There's nothing better than seeing my idea catch the attention of capital market veterans. We will forever change the course of finance.

What dreams do you have for the next 10 years?
Seeing technology get to market, and then the commercialization and adoption of it. Then, taking the model and using it for another vertical: wash, rinse, and repeat.

If you could change one thing about the world we live in, what would it be?
The pace at which we do business is lightning fast and we are bombarded with so many input streams. I would like to have all interactions focused without interruption.

What was your first technology investment and what happened?
It was a great idea that took too long to develop into a business. While the investment was modest, it still stung. Coming from a big corporate background, I learned the lesson of good vs. perfect. Getting to market was more critical than fully baking every detail.

Can you share a story of one of your best investments, and why it went so well?
The best investment had an early exit…HOORAY! The CEO was humble and accepted coaching. The team was lean and all the members worked well together. More companies die from suicide than homicide; corporate culture trumps strategy any day.

Can you share a story of one of your worst investments and why it went so badly?
There was a fight between 2 of the founders and a power struggle ensued. The argument became so vitriolic that it collapsed the business.

In your opinion, what is so special about Silicon Valley?
Silicon Valley is a place where anything is possible. We don't say "you should do something"; we look at things and say "we can tackle that". Also, there's a genuine sharing of ideas and willingness to open your network/connections to help advance projects in way that doesn't exist in other locations. As a New Yorker, and having spent most of my time on Wall Street, it's a refreshing atmosphere and attitude.

Why do you think the innovation economy/ Silicon Valley has a poor track record on diversity and inclusion? What needs to change?

As a female CEO and founder, in 2019 I had a meeting with a male VC and one of my male colleagues. The VC did not look at me or ask me a single question in the meeting. It was so palpable that my male colleague apologized to me after the meeting concluded. There is no excuse for it. At my company, we hire the best person for the job regardless of gender, ethnicity or sexual orientation. We have a fairly even split between men and women. It's not hard, it just needs to be top of mind. There also needs to a culture where diverse perspectives are respectfully and professionally heard.

There is a lot more focus now by investors, employees and customers on companies that do good (and do no harm) as well as make money. How do you think about this topic in the context of your investing?

You can conduct business in a respectful and professional way that also does good. It may mean that you turn down money from VCs without a female or ethnically diverse team. Further, I will not invest in companies that do not have diversity in their team. Doing good takes more time but it's worth the extra effort.

Who have been role models you most admire?

Meg Whitman—I had the chance to meet her last summer in Aspen. I adore the fact that she is creating yet another company in a new focus area (going from a series of career steps at amazing companies to eBay to HP to now Quibi with Jeffrey Katzenberg). Taking eBay from $4M to $8B in revenue is truly remarkable. She's always reinventing herself.

What is the best advice you ever received about how to succeed in Silicon Valley?

Do the work. I'm known for GSD and it's important to do what you say you are going to do.

What do you believe are the most important traits of a technology entrepreneur?

Drive and resilience. Setbacks are inevitable but it's all how you relate and overcome hurdles and obstacles that sets you apart from those entrepreneurs who don't make it.

What do you believe are the most important traits of an early stage technology investor?

Careful research and talking with end clients. There is nothing second to field research and knowing the mindset of the customer.

If you could tell a technology entrepreneur just one thing, what would it be?

Keep at it. There is no magic or substitute for hard work.

What do you know now that you wish you had known earlier on?

How hard it is as a female to raise money.

Would you please create a personal quote that captures important wisdom about participating in early stage investing?

"Early stage investing is all about relationships and diversification. Do your homework in learning about the team. There is always a point in due diligence (about 40 hours into the DD) when the true character of the entrepreneur and their team come through."

Imagination, inquisitiveness, and humbleness. The best meetings with entrepreneurs are when they prove you totally wrong and dispel your misconceptions. That's how you know that they really know their stuff.

"Invest in deals where you understand the product and the market fit but expect that both will continue to continue to evolve, so the agility of the team to iterate rapidly will dictate the success or failure of the company."

Claudia Fan Munce

Angel Investor, Former Venture Capital Partner, Advisor

Broadway Angels, NEA and Board Director Best Buy, Corelogic, Bank of the West

What did you want to be as a child?
Engineer.

What was your first career?
Computer Science Engineer at IBM Research Laboratory.

What other careers have you had before your current one?
30 years with IBM, starting as an researcher, head of Licensing and Commercialization, then head of Venture Capital Group and Corporate VP of M&A.

Why do you choose to focus on early stage technology investing?
As a strategic investor, early stage companies represent a pipeline of innovations that help large corporations build the M&A and partnership pipelines.

When your investing is not going well, what do you do or say to yourself to keep motivated?
The company with a good team can rapidly iterate its strategy and product to adapt to better serve the market or even adjust its target market. Good companies are able to adapt and evolve into a successful company with the appropriate guidance.

What do you love most about what you do now?
Innovation is exciting and in time like this, it is critical as technological innovation will play a critical role in defining our future, the way we live, the way we work, the way we learn - every aspect of our lives.

What dreams do you have for the next 10 years?
Equality of opportunities for all. As a poor immigrant growing up in Brazil, I came as a foreign student on scholarship to U.S.A, and had a successful life filled with great opportunities that afforded me with wealth and security. My dreams is that the same opportunity that afforded me 30 plus years ago will continue to be available to all those who are willing to work hard to succeed. "The American Dream", once inspired people from around the world including a poor student from Brazil, seems to be dissipating even for our own children born here in the U.S.A.

My dream is for this country to return to be the beacon of light to the world instead of the darkness that we are all living in right now.

If you could change one thing about the world we live in, what would it be?
Get rid of Donald Trump and all his dishonest associated from any position of power that they can do damage to other people, especially the vulnerable ones. With all the crisis, pandemic, climate, economic, poverty...there is nothing more critical than restoring the humanity in this country.

What was your first technology investment and what happened?
It was year 2000 before the Internet bubble, the company created a thin server that can pack in server farms in a fraction of real estate with innovative head dissipation technology (cooling). Anticipating that Internet will break all projected CPU, storage and networking capacity provisioned by our big financial services customers, I lead a significant investment in a new company that will transform on premise datacenters density and unleash unforeseen capacity. Internet bubble burst and capacity demand collapse and company went out of business. Great technology, great team, but the macroeconomics took the company down as investors fear for the uncertainty and were not willing to inkjet new capital needed.

Can you share a story of one of your best investments, and why it went so well?

As a strategic investor, we wanted to roll out a new Machine Learning platform and was looking for a startup that can build on our platform to demonstrate the capability of the platform and serve as a strong reference use case for the platform. I was able to select from the top tier venture backed companies with very strong investors base to syndicate a new round to enable the company to scale on our platform and broaden the market that it serves. A successful company today. The fact we knew exactly what we wanted to get out of the investment and we were clear on the value we can provide to the company allowed the deal to be very successful. It is rare to have that level of clarity going into any deal.

Can you share a story of one of your worst investments and why it went so badly?

A company we invested because we want to acquire the capability that it was developing. We did acquire but horrible founders team with complete cultural mismatch resulted in great failure in integration and eventually we abandon the deal and release the team.

In your opinion, what is so special about Silicon Valley?

Rapid iteration as a model of innovation, not focus on pushing the envelope of physics but innovation to remove friction in jobs that we need to do as part of our lives. Allowing failure to happen fast, therefore allowing the learning to be injected into the evolution of the next generation of products and services, in rapid speed. Silicon Valley attracts people that are willing to fail in order to succeed, and those who fail are rewarded with respect and very often multiple more chances to succeed.

To be amongst the intellectually curious and driven people is exciting if you are one of them, it is a community connected by its inspirations and everyone has great hope to make an impact (and make money as well) The Valley attract the bests of the best from around the world, and demands respect from the world as a hub of innovation responsible for transformational paradigm shift in the way we live, injected by the technology created here but servicing the world.

Why do you think the innovation economy/ Silicon Valley has a poor track record on diversity and inclusion? What needs to change?

I don't think it is on purpose a discriminatory gesture as I do feel it is human nature to want to work with people that you can identify with. The early investors to Silicon Valley are all white men coming from East Coast and clearly they started this Valley with their hard work and wisdom and recruited people that they can identify with. Clearly we are catching up top a long history of this but if you look at the Indian entrepreneur and investors, even CEOs, I would say they are no longer in minority, once they reached a critical mass and people recognize them as winners, the same behavior persisted amongst them and result is a dominant community in Silicon Valley composed of Indian or Indian descendants.

There is a lot more focus now by investors, employees and customers on companies that do good (and do no harm) as well as make money. How do you think about this topic in the context of your investing?

Very important topic as we all need to contribute to reversing the inequality that exist in our society. If we don't factor the problems we are trying to fix in our decision making process, we will never be able to improve the problems we face, and they will eventually hurt all of us. So to make an explicit effort to balance all the metrics you impose in your investment decisions is a way to contribute to a better future to us all.

Who have been role models you most admire?

My mom, grace under fire. She had a hard life but never lost her grace. She shows up with confidence everywhere she goes even when she has no right to be there. She was able to nurture her family through a lot of hardship and left behind 4 very successful children that are all extremely grateful for her love.

What is the best advice you ever received about how to succeed in Silicon Valley?

To nurture key relationships from people that you can add value to and therefore will return the value back to you.

What do you believe are the most important traits of a technology entrepreneur?

Physical and emotional stamina, intellectual curiosity, creativity and humility. The successful entrepreneurs have broad sensory radars himself/herself or surround themselves with people that are part of their sensors. They listen carefully to all insights and synthesize them to make best decisions in a continuous loop.

What do you believe are the most important traits of an early stage technology investor?

Risk prompt personality, ability to trust others, and excited about innovation and disruptions.

If you could tell a technology entrepreneur just one thing, what would it be?

Do your homework always, never show up unprepared, you don't get many chances to impress and your knowledge about who you are talking to and why you are talking to them is critical to lead to a second conversation.

What do you know now that you wish you had known earlier on?

That arrogance is inversely proportional to the caliber of both entrepreneurs and investors. Walk away the moment you find a person arrogant.

Would you please create a personal quote that captures important wisdom about participating in early stage investing?

"Invest in deals where you understand the product and the market fit but expect that both will continue to continue to evolve, so the agility of the team to iterate rapidly will dictate the success or failure of the company."

> *"You only have to do a very few things right in your life so long as you don't do too many things wrong."*

Lilianna Nordbakk

Angel Investor, Founder, CEO

Band of Angels, Life Sciences Angels, TréBiome Inc.

What did you want to be as a child?

As a child, I was very curious about everything in nature. My grandfather gave me a microscope when I was eight years old, and I remember studying anything I could find in nature. From then on it was so clear to me that I wanted to be a medical doctor or scientist. I wanted to heal people and alleviate suffering. Unfortunately, calculus got in my way. In the German school system at the time, you needed a straight A to go to medical school, and I fell short. It was brutal. But I never lost my love for science. After my retirement as an executive, I went back to school to study psychology and biotechnology. Studying psychology and biotechnology is how I first learned about microbiome medicine, which is now my new passion area. Your childhood dreams can be revived!

What was your first career?

Language interpreter. It actually helped me to become a more successful entrepreneur and leader because it taught me how to carefully listen. It's a very important skill set to build as an entrepreneur.

What other careers have you had before your current one?

Serial entrepreneur with over 20 years of experience in business development and finance in the software industry. Co-founded several companies in Europe and ran them as a CFO from inception to an IPO or acquisition. This includes NorCom Information Technology, which is a leader in AI for the automotive industry in Germany. I have been an Angel Investor since 2005. Most recently, I started TréBiome Inc, which incubates and invests in microbiome-centric startups.

Why do you choose to focus on early stage technology investing?

Despite the risk, this stage is most rewarding – both personally and professionally. Unlike other investment types, I can investment my funds AND my entrepreneurial experience.

When your investing is not going well, what do you do or say to yourself to keep motivated?

Not every investment will be successful, so I focus on the ones going well. These make up for the rest.

What do you love most about what you do now?

Through angel investing, I meet people with fascinating ideas. And investing in life science startups is my way to support what I care about – improving people's lives by combating diseases.

What dreams do you have for the next 10 years?

Personally, I want to live on a farm, grow food, keep animals, and enjoy nature. Professionally, my dream is to run an angel investment fund specializing in microbiome medicine. The human microbiome is a community of microorganisms that live in or on the human body. An exponentially increasing body of scientific knowledge has demonstrated that these microbial communities perform a wide variety of important biological functions and thus can be harnessed as therapeutics to treat diseases with high unmet medical needs.

If you could change one thing about the world we live in, what would it be?

If I could change one thing about the world, it would be the way we deal with climate change. Climate change is the most pressing threat humanity has ever faced. Changes in the natural balance of the earth's atmosphere, caused by human behavior, are having a significant impact on the environment. Scientists from all over the world agree that the effects of climate change will get worse unless we take action now. Because the actions of countries cause climate change, it can only be solved by the cooperation of countries, -- and that means international law offers the best pathway forward.

What was your first technology investment and what happened?

During the '80s, client-server computing disrupted the software industry by using a distributed computing model. In the new model, processing power was located within small, inexpensive client computers linked to one another through a network. It replaced centralized mainframe computing, where nearly all of the processing took place on a central, large mainframe computer. IBM, who dominated the marketplace at that time, wasn't ready for the new software paradigm. Hence, many new technology startups had a chance to enter the market and challenge enterprise software solutions.

I was in my early twenties when I had the opportunity to co-found one of these software companies in Germany and to invest my savings to get founder shares in the company. The lack of venture capital or alternative sources for risk capital forced us to bootstrap the company growing it from two to 400 employees over 10 years. This was very challenging because not having enough funding slowed down the company's growth. Another drawback of bootstrapping was the lack of investors who acted as mentors and who could offer a network of valuable industry contacts. However, through bootstrapping I was able to keep all my founder shares until we took the company public, and therefore this investment turned out to be the very best one so far from a return perspective.

Can you share a story of one of your worst investments and why it went so badly?

I made a few investments that didn't go well. These investments shared one commonality centered around the leadership's management style. Each startup had strong-minded leaders that were unwilling to listen to the market and pivot based on what they learned. It's important for entrepreneurs to balance their determination with being humble to the market.

In your opinion, what is so special about Silicon Valley?

Silicon Valley is a fertile ground for startups, where everything a startup needs to be successful is available and concentrated in an area that spans just about 20 miles. With Stanford University and Berkeley, there is a constant flow of talent. There are thousands of wealthy investors in the area willing to write a check for a great idea. In addition, these investors often act as mentors and provide a transfer of entrepreneurial wisdom from one generation of entrepreneurs to the next. Lastly, the United States offers startups with a big domestic market, which is advantageous if your goal is to build your company into a global market leader. In Europe, for example, there are more barriers to entry when expanding your business, such as language and culture.

Who have been role models you most admire?

Ruth Bader Ginsburg is inspiring. She's a warrior for women's rights, using her voice to advocate for justice. She's purpose driven. When setting priorities, I think of her advice: "How do I add value to my organization and society? Do I have a purpose that motivates me to be my best self?"

What is the best advice you ever received about how to succeed in Silicon Valley?

The best advice I ever received was to *"Never, never, never give up no matter what"* This is a mantra that I live by and has helped me keep pressing forward in business, even when things don't go well. There will be problems. That's inevitable. I experienced many when growing my companies. But know that if you simply "stay in the game" and don't give up, you will be successful after all.

What do you believe are the most important traits of a technology entrepreneur?

Vision.

Passion.

Pragmatism.

Problem-solving.

Flexibility.

Bravery.

Self-discipline.

What do you believe are the most important traits of an early stage technology investor?

The most important character traits for an early-stage investor are integrity and collegiality to foster a trustful relationship with the entrepreneur. The investor must be willing to roll up his or her sleeves and be hands on when it comes to solving problems *together*. It is a great plus if the investor knows the sector space very well and has a network that can provide support to the entrepreneur.

If you could tell a technology entrepreneur just one thing, what would it be?

I would give the entrepreneur the same advice I once received: manage your attitude and never give up, even in the darkest of times. Never give up and learn from your mistakes.

What do you know now that you wish you had known earlier on?

To view losses as educational experiences.

Would you please create a personal quote that captures important wisdom about participating in early stage investing?

"You only have to do a very few things right in your life so long as you don't do too many things wrong."

This is my favorite Buffett quote. In other words, long-term success belongs to those early-stage investors who consistently focus on doing a few things right, rather than focusing on avoiding all mistakes. Focusing on mistakes will frustrate and paralyze you. Accept that you're going to make mistakes and not all investments will pan out. It's part of the early-stage investing process.

"Listen to what's being said, not said & trust your instinct only as much as you can afford to lose."

Karim Nurani

Angel Investor

Sandalwood Ventures, Linqto

What did you want to be as a child?
As a child I wanted to be a Millionaire, Explorer, Rally Driver, Safari Guide in the African Bush.

What was your first career?
My first Career in the equity / investment world was the General Manager of a gold mining company in Tanzania in the early 90's. I was just out of college and met up with a Canadian mining company that was setting up a gold mine in Tanzania they were raising funds but since I didn't have any I took up this role as my equity participation.

What other careers have you had before your current one?
Prior to arriving in the US in the early 2000's I was the general manager for a steel rolling mill in Kenya and continued my journey by building & operating the first TexMex restaurant, the first Wrapps & Smoothie restaurant & finally the first automated American style bowling alley. In the US I started my sales career with a financial company & subsequently worked in enterprise sales for two publicly traded companies selling HR outsourcing services.

Why do you choose to focus on early stage technology investing?
I actually did not choose to focus on early stage technology investing, I chose to assist entrepreneurs who were building exciting companies that were addressing unique needs in the marketplace that this is coincidental with developing unique technologies was interesting to observe.

When your investing is not going well, what do you do or say to yourself to keep motivated?
Ha Ha, most investments have a rocky path to travel & if you decide to console yourself because you have made an investment that's going south you are in trouble. Rather know that there are going to be more bad days than good so focus on developing a mental & spiritual stamina that will see you through these rough patches.

What do you love most about what you do now?
I enjoy meeting entrepreneurs that are excited about what they are building & I enjoy helping them build their company. Oftentimes these entrepreneurs need to build & manage their teams or explore sales opportunities or need mentorship guidance & these are what I enjoy doing the most.

What dreams do you have for the next 10 years?
The first ongoing dream is to stay with this great startup today. At Linqto we are building a platform & creating a market place that allows accredited investors to invest in pre-IPO companies simply & affordably. An investment area that has been impossible to access unless you are already a millionaire. I believe we are in the right place, at the right time with the right product. The dream continues with me building a health & wellness retreat on the Coast of California or Southern Europe. Where personal boxes remain unchecked & individuals allow their true characters to expand & flourish & the simplicity of life can be truly appreciated.

If you could change one thing about the world we live in, what would it be?
The world needs to support the children of the future by providing food, shelter & education. There are too many children in this world today that are fighting emotionally & physically just to survive & this does not allow them the opportunity to live a fulfilling life. Once the youth are embroiled in this cycle they are diverted from focusing on social justice, environmental degradation, political & economic corruption.

What was your first technology investment and what happened?
My first investment in the US was with a company that created & patented the technology to allow for a unique non-moving parts dispensing kiosk. This allowed the US to deploy a highly superior retail kiosk, which cost less to manufacture, cheaper to maintain, had a much smaller footprint so we could deploy without space constraints. Since we built this

as an end-to-end solution we also developed a better supply chain & data tracking software system. This allowed us to track products, revenue & margin with real time data. We also developed a refillable cartridge system that allowed us to manage inventory to a just in time basis that allowed us to control expenses & drive profit with a data driven insights. We sold the technology to a much larger company that was the original leader in kiosk retail business.

Can you share a story of one of your best investments, and why it went so well?

A few years after graduation from college in Canada I was back in Africa working for an International bank learning about Forex exchanges, I met a group of Canadian geologists who acquired a gold mining license in the remotest part of the newly liberated country on Tanzania in East Africa. They were raising funds to develop the mines. My investment came though as sweat equity. I joined as general manager with in country experience who was familiar with cultural nuance & language.

We dug wells & dams, built roads & mud huts, slept in tents played tag with scorpions were constantly dodging king cobra attacks or black mambas hiding in shrubs while we hiked our claims. We survived on peanuts & bananas fought our way through malaria & dysentery.

We imported ore crushers from India, land cruisers from Japan, and shotguns from Canada. So many things went wrong, import / export licenses monsoon rains, equipment failure thefts, floods, deaths & illness. Two years later we smelted our first gold bar drove that to the bank in a convoy of land cruisers riding shotgun. The most exciting experiences I have ever had. But by this time we were raising funds only to survive and we finally were completely diluted by follow on investors.

Can you share a story of one of your worst investments and why it went so badly?

Invested in a great opportunity within the FinTech space. The founder & CEO was newly relocated to Silicon Valley, he had been successful in the past, had a great personality & was an inspirational team leader. Unfortunately, he got so caught up in the Silicon Valley Hype of growing the company fast that they played fast & loose with their investors funds & reporting of financials and this finally caught up to them & were indicted.

In your opinion, what is so special about Silicon Valley?

The energy & enthusiasm & culture here is so infectious that it makes you believe that you can achieve anything you set your mind too. The same culture also allows you to fail & will not punish you for that as long as you learn very quickly.

What is the best advice you ever received about how to succeed in Silicon Valley?

Understand who you truly are, what is important & keep your moral compass. In Silicon Valley there is so much hype on success & the image of success that sometimes it is easier to take shortcuts.

What do you believe are the most important traits of a technology entrepreneur?

Understand your strengths acknowledge your weakness then try to identify team members that are exceptional at what they do & trust them to do what is needed.

What do you believe are the most important traits of an early stage technology investor?

Invest in an area that you are most interested in & have experience in that landscape. Often times entrepreneurs value your knowledge & connections much more than the money. Be honest & open to entrepreneurs if you're not interested let them know early & provide a reason. This helps them understand your focus better.

If you could tell a technology entrepreneur just one thing, what would it be?

Identify investors that have experience, connections & passion about who you are and what you are trying to achieve. Try to get to the 'no' as quickly as possible without being arrogant.

What do you know now that you wish you had known earlier on?

We will never know everything but the key to not to forget what we have learnt in the past. Research & due diligence on the team background & internal team dynamics is really important. How well do they work, communicate & resolve problems.

Would you please create a personal quote that captures important wisdom about participating in early stage investing?

"Listen to what's being said, not said & trust your instinct only as much as you can afford to lose."

Having an excuse to constantly meet, learn from, and be inspired by people much smarter than myself.

"Early stage technology investing is still so new…now, it is capturing the public's imagination, entering the zeitgeist, and becoming rapidly more democratized."

Alex Pack

Venture Capital Partner

Dragonfly Capital

What did you want to be as a child?
Spy.

What was your first career?
Lecturer at a small liberal arts college in Hong Kong, teaching English language and literature.

What other careers have you had before your current one?
Founded an education technology startup. Built crowdfunding products at AngelList, the largest platform for early stage tech fundraising.

Why do you choose to focus on early stage technology investing?
The only way to see the future is to invent it yourself - but financing and supporting the inventors is a close second.

When your investing is not going well, what do you do or say to yourself to keep motivated?
In the micro, startups are fragile and prone to failure. But in the macro and over the long term, they have to succeed.

What do you love most about what you do now?
Getting to speak with and support incredibly smart people at the frontiers of their fields every day.

What dreams do you have for the next 10 years?
I'm most interested in civilizational-level shifts. In crypto and FinTech, my current focus, we have the potential to totally transform the financial system, which mostly hasn't changed in over a century, in a digital-first, open and transparent way. Finance underscores the entire economy, including my own industry of venture capital, and there's so much to fix over the coming decades.

If you could change one thing about the world we live in, what would it be?
Growing wealth inequality is the biggest social issue in the Western world today. Yet it's clear that inequality is a political problem and technology alone cannot be the solution. In a decentralizing world that is getting more and more fractured, grand political solutions seem less likely but are more needed than ever.

What was your first technology investment and what happened?
I started in venture as an analyst at a FinTech venture capital firm in Hong Kong. Since I was an American, I focused on cross-border deals, and since I was the youngest, I volunteered to be the blockchain lead. I spent a year diving into the space before we made our first investment, meeting hundreds of teams and peppering now-unicorn company founders with the most basic questions. While our first investment is still alive and doing fine today, we passed on several multi-billion-dollar companies and multi-ten-billion-dollar tokens. Ultimately, it was a problem of imagination. Even though we were disruptive technology investors, there was still an image in my head of what a typical venture-backable startup should look like: an equity investment into a company, a founder with prestigious tech or exec-level domain expertise, a clearly explainable use case for mainstream users. I would have been better off investing in the exact opposite: tokens and odd structures, founders off the beaten path, and either infrastructure platforms or use cases that barely look like use cases. I was looking for blockchain-powered remittance applications when I should have been thinking about new types of *money* and value-transfer platforms – the things five levels deeper.

Can you share a story of one of your best investments, and why it went so well?
Several years later I joined Bain Capital Ventures and helped kick start their crypto investing. At this point, crypto had become more mature and it seemed like we were finally ready to see some of the early end-user applications that I was expecting when I first entered the space. Especially in decentralized finance, which was a massive opportunity: using the blockchain to allow anyone in the world with an Internet connection to access, create, and exchange their own financial products. The first companies I backed had tried going after these use cases years

ago, but there simply wasn't enough infrastructure to make it possible. I met the CEO of this new company very early. Not only did he not have a working product, he didn't have a pitch deck. He architected out his initial product design on a whiteboard and his most frequent answer to my questions was, "I don't know, but it will be really fun to figure that out." His plan was to build the most complex decentralized financial application yet, then eventually launch a token and dissolve the company. We gave him a few million dollars, became the largest investor in his seed round, and did our best to support him along the way. Today it is one of if not the largest decentralized applications in crypto.

In your opinion, what is so special about Silicon Valley?

I started my venture career in Hong Kong and I've since invested in companies around the world – Beijing, Shanghai, Singapore, Berlin, Tel Aviv, and all over the U.S. But there's something that has always made Silicon Valley special. It is by far the most diverse city of tech founders I've backed. Most are immigrants or the children of immigrants, they have the most unusual backgrounds, and they've come here from around the world with a single purpose.

Entrepreneurialism is not evenly distributed across the population. It is concentrated in the tails, in a small number of eccentric risk-takers. So it doesn't really matter how Silicon Valley became the startup mecca, it's the fact that a mecca exists at all for would-be entrepreneurs to flock to that is important.

Things have shifted in recent years: rising cost-of-living at the local level, tougher immigration at the national level, and de-globalization at the global level. We will see if Silicon Valley can survive these trends, but if it doesn't, it's unlikely that there will ever be one single 'startup mecca' again for many years.

Who have been role models you most admire?

Alexander Hamilton, Peter Thiel, Henry Kissinger, Karl Marx, Joseph Schumpeter. Thinkers who have been able transform their ideas into the sphere of action. As Marx said about his own contemporaries, *"Philosophers have hitherto only interpreted the world in various ways; the point however is to change it."*

What is the best advice you ever received about how to succeed in Silicon Valley?

Emotion is the mind-killer. You lose and get put in difficult situations all the time, but Silicon Valley is a long-term, repeat game with recurring participants. Best not to take things personally.

What do you believe are the most important traits of a technology entrepreneur?

Imagination, courage, and unreasonable determination

What do you believe are the most important traits of an early stage technology investor?

Imagination, inquisitiveness, and humbleness. The best meetings with entrepreneurs are when they prove you totally wrong and dispel your misconceptions. That's how you know that they really know their stuff.

If you could tell a technology entrepreneur just one thing, what would it be?

I don't think we appreciate entrepreneurs enough as a society, even today. It's an all-consuming life that often results in failure for many years. Founders are the heroes of our post-political era; they drive the world forward against all odds.

Would you please create a personal quote that captures important wisdom about participating in early stage investing?

"Early stage technology investing is still so new. Historically, it is probably the last field of finance to develop into its own industry with best practices: centuries passed before its practitioners were able to make it systematically profitable and sustainable. Now, it is capturing the public's imagination, entering the zeitgeist, and becoming rapidly more democratized. The key is to remember that we are still in the early innings, with so much change ahead of us."

I believe in a world where entrepreneurship becomes a true meritocracy, where women and other underrepresented groups have a level playing field.

"For me, it starts with respecting the sacrifices and tenacity of the founders I work with and then investing in companies where I believe we can add value well beyond the capital that we provide."

Cindy Padnos

Venture Capital Partner and Angel Investor

Illuminate Ventures and J Angels

What did you want to be as a child?
In a 6th grade essay, I said that I wanted to be an equestrian with my own professional barn, training and riding Olympic-level horses. I also said that I would buy my dad a schooner and my mother a mink coat. Poor Mom would have gotten the raw end of that deal.

What was your first career?
My first job after grad school was in high-tech sales, but I never thought of sales as a career. It was just a great point of entry. I suppose the career part was "B2B/enterprise tech". While I have played many different roles in that category, I have never strayed far.

What other careers have you had before your current one?
Operating exec in high tech – both large corporate and startups, Management Consultant, Founder & CEO of a high-tech company.

Why do you choose to focus on early stage technology investing?
As a prior tech entrepreneur, early stage investing offers the opportunity to learn something new from creative founders every day and in parallel to leverage my prior experience to add value.

When your investing is not going well, what do you do or say to yourself to keep motivated?
The path to success is rarely straight up and to the right. Bumps are to be expected.

What do you love most about what you do now?
It's all about the people. I've had the privilege to collaborate with several founders over many year periods as they grew their companies from zero to over $100M in revenues.

What dreams do you have for the next 10 years?
I believe in a world where entrepreneurship becomes a true meritocracy, where women and other underrepresented groups have a level playing field.

If you could change one thing about the world we live in, what would it be?
I would ensure that every child has access to a great education. I believe that education from infancy onward is one of the most important keys to enabling success in life and to creating a more open and tolerant world.

What was your first technology investment and what happened?
I had done several successful angel investments, but the first true VC ($1 million) investment I made was an utter disaster. I was at a traditional early stage firm and followed another well-respected partner who was leading a Series A financing of a Grid computing platform. Unfortunately, 12 months later, after several millions of dollars had been spent, we came to understand that the technology was simply not nearly as good as it had looked on paper. The only good thing about it was that I didn't know that it would fail until after I had already made three other investments that turned out to have very positive outcomes (one IPO and 2 significant M&A exits). Had I known the first one would go on to fail, I might have thought that I just wasn't going to be a good investor and stopped there. Fortunately, that didn't happen. There were several lessons learned: It was a great illustration of why a portfolio of companies is important and also the value of investing in areas that you are familiar with and can value to. In the end, at the founder's request, we sold the company back to him for the cost of the legal transaction fees. He never had any success with the company, but we didn't wish to stand in his way if he wanted to continue to try. Fortunately, it was another 10 years and two dozens investments before I have to live through that kind of a company implosion again.

Can you share a story of one of your best investments, and why it went so well?

I have been fortunate to have several very successful investments, but none of them happen overnight. I invest only in the B2B/Enterprise software space, where the average timeframe to exit is 7-10 years. One of the terrific companies I invested in is Xactly Corporation. It was an 11-year process from initial investment to the ultimate outcome, a greater than half-billion-dollar acquisition by Vista, with a successful IPO (XTLY) along the way. I was a board member of the company for 10 of those years, so had a close-up view of the company's evolution. The company started with a few key fundamentals in their favor. The founders had deep domain experience, they were the first SaaS offering in the category, so they could target a larger opportunity set than traditional on-premise products and the CEO himself was a "selling machine". Amongst the things the company proactively did right was their decision to implement an internal mandate, from day one, to focus on people. The company had been founded by a very diverse team (Latino CEO, CTO from India and female head of products) which understood the importance of building a great team and also giving back to the community. The company was rewarded for these efforts by gaining and holding the Fortune "Best Places to Work" awards for many years– which aided their recruiting and retention efforts. I have to admit that I actually had tears in my eyes as I stood on the floor of the New York Stock Exchange and watched the original founding team (and many others) ring the opening bell for their IPO.

In your opinion, what is so special about Silicon Valley?

Silicon Valley is unique in many ways, but one thing that sticks out for me after traveling the world is how the concept of entrepreneurship is looked at very differently here. It's not just that founders are celebrated here, it's that we even admire them when they fail, as long as they did their best and learned from the process. I remember the first time I went to Australia to look at a possible investment and several people there told me that the perception of an entrepreneur there was someone who started a company because they couldn't get a job! Granted, that was 20 years ago, but I believe Silicon Valley continues to stand far ahead of most other geographies in celebrating the combination of risk taking and visionary thinking. Even my hair stylist thinks entrepreneurially.

Why do you think the innovation economy/ Silicon Valley has a poor track record on diversity and inclusion? What needs to change?

Unintentional bias is the biggest obstacle to diversity and inclusion in Silicon Valley and throughout the world. It's nearly impossible for most people to recognize something that they do not feel they could possibly represent. They just don't see it in themselves, so they never even attempt to address it.

Back in 2010 when I decided to do some research about why women were gaining so little VC funding (less than 5% at that time), I learned about the concept of homophily (a natural affinity that we all have to be more comfortable with people who are more like ourselves) and found a vast amount of data showing that women faced much higher barriers than men in gaining access to capital. I ended up writing a white paper on the topic (*High Performance Entrepreneurs: Women in High Tech*) that shared a "glass half full" view of how those obstacles were mostly in place due to myths that simply did not any longer hold true (like that women were risk adverse or that you had to be an engineer to be a tech founder). There wasn't enough data available to compare the results with other diverse groups. There were and still are just too few. Not all that much has changed 10 years later. Our 2018 research (Gender Differences in Entrepreneurship: Voices of Founder and Funders) found that that both male and female investors still demonstrate significant unintentional bias regarding female founders.

There is a lot more focus now by investors, employees and customers on companies that do good (and do no harm) as well as make money. How do you think about this topic in the context of your investing?

As a firm, Illuminate does not have a specific mandate around diversity or "doing good", but we do go out of our way to find ways to level the playing field for all entrepreneurs and to ensure that the companies we invest in do not unintentionally have a negative impact, whether that be socially or environmentally. As an example, we have run training sessions for our portfolio companies on how to expand diversity within their companies. In one instance we brought in a consultant who had worked with Facebook and other large corporations to share best practices. One thing we know for sure is that establishing diversity early on makes it much easier for a company to succeed in that area as they grow. As an early stage investor, we feel that we have the privilege and a responsibility to help influence that. We see this as both good business and simply the right thing to do.

Throughout my investing career with Illuminate, I have also had the opportunity to build a Business Advisory Council made up primarily of corporate execs and entrepreneurs in the B2B/tech ecosystem. It was always my intent to create a very diverse group to support not just our investing team, but also to offer access to our founders. The current group of over 40 members is more that 60% female and three fourths of the members are diverse – meaning non-white males. They are an amazing resource to our partnership and to our portfolio. Many of them are involved as a means to give back into our entrepreneurial community, and we work hard to ensure that they have this opportunity.

Who have been role models you most admire?

My father was an amazing role model. An entrepreneur, born to immigrant parents he worked nearly full-time into his 80s, sailed into his early 90s, and taught me the importance of integrity. In the Venture community people like Rory O'Driscoll, Tom Bredt, Nancy Schoendorf and Cliff Higgerson each also taught me things that have helped me become a better investor.

What is the best advice you ever received about how to succeed in Silicon Valley?

Early on as a founder/CEO one of my investors urged me to "never settle" when hiring an early team member. He said, "you only get one of those this year – take your time to find the best person". It was great advice, but not easy to implement during the "bubble" years when there was literally a war for talent. I found out, painfully, how smart the advice was when I did in fact "settle" for a candidate I had concerns about. There are few things that have more impact in life than the people that you surround yourself with, whether that be as an operator or as an investor.

What do you believe are the most important traits of a technology entrepreneur?

Resilience and inspirational leadership are two traits that I look for in entrepreneurs. When we did some research recently on this topic across several hundred entrepreneurs and investors, resilience/perseverance was ranked uniformly high by all of the participants. The ability to attract a great team and relevant domain knowledge/expertise were higher ranked by investors than by founders and the founders ranked having a vision/passion for the business as being of higher importance than did the investors.

What do you believe are the most important traits of an early stage technology investor?

Intellectual curiosity - basically a willingness to listen and learn, is one of the most important attributes of an early stage investor. At Illuminate we are always looking for out of the box, new ideas and new types of founders. We do our best to avoid typical pattern recognition that might inhibit the likelihood that we explore new concepts that may feel a bit foreign at first. This open mindeness needs to be paired with both intellectual and business integrity. It is critical for a founder to know they can trust you.

If you could tell a technology entrepreneur just one thing, what would it be?

My advice is simple: 1. Be sure that you are truly passionate about what you are doing 2. Spend as much time developing an understanding of your customer's needs as you do on building the product (or more) and 3. Surround yourself with people (employees, advisors, investors) who you honestly believe are smarter than you are – at least in some areas.

What do you know now that you wish you had known earlier on?

When I was a first time CEO, I honestly did not fully understand that while I worked for the board, that I actually needed to manage them and their expectations. It feels like a rookie mistake now, but I didn't have a group of other CEO friends at the time to learn from. When I made the transition to become an investor, I did not have a full appreciation for the time and effort required to raise capital, particularly as a first-time fund manager and a woman. Again, I did not have many peers to reach out to and learn from. I think the lesson is to not feel limited by your own network – to take the time to gain access to resources that can help.

Would you please create a personal quote that captures important wisdom about participating in early stage investing?

"I think about Seed and early stage investor as a privilege, not a right just because you have capital to deploy. For me, it starts with respecting the sacrifices and tenacity of the founders I work with and then investing in companies where I believe we can add value well beyond the capital that we provide."

> *"It's fine to be long term but even marathon runners watch their early pace."*

Stephen Parlett

Angel Investor following 20+ years in public Technology, Media and Internet funds

Keiretsu Forum, Linqto, Blockchain Coinvestors

What did you want to be as a child?

A businessperson. At University, when I studied Investments and Mutual Funds I knew I wanted to target a career in that field.

What was your first career?

Portfolio Accountant for G.T. Capital which was a fast growing firm specializing in international and emerging market mutual funds.

Why do you choose to focus on early stage technology investing?

Given my experience in working with C-suite executives of TMT companies and doing extensive qualitative and quantitative research, I now have the opportunity to apply that to earlier stage investing.

When your investing is not going well, what do you do or say to yourself to keep motivated?

"I did not become dumb overnight." "Trust the process and continue to be objective about the inputs and results."

What do you love most about what you do now?

I am now fully dedicated to my personal portfolio. I am both leveraging my prior investment work and broadening my horizon to include private investments.

What dreams do you have for the next 10 years?

For myself, it is to take my prior professional skills and apply them to improve organizations both profit and non-profit. With high school aged kids, I want to be fully available to help guide them through high school, university and their first career.

If you could change one thing about the world we live in, what would it be?

I would like to turn everyone into an objective and independent thinker. I believe we could be a much more functional society if we looked at each issue free of historical bias and prior affiliation. Each social, economic and environmental issue should be assessed and implemented on the facts and modeling of multiple scenarios that take into consideration short, medium and long term impacts as well as unintended consequences.

What was your first technology investment and what happened?

I was managing a TMT focused mutual fund in the late '90s. We were early on the research of DSL technology and regulation that would enable multiple new companies to offer higher speed Internet services to businesses and homes utilizing existing telecom infrastructure. Covad Communications was one of them. Ultimately, the monopoly telecom company was threatened and had more political and operational weapons to fight back. The result was slower adoption curves and a poor customer experience that lasted longer than the company could overcome before coming profitable. That was one example of many early lessons to be learned from investing in an early stage company despite it being publicly traded at the time.

Can you share a story of one of your best investments, and why it went so well?

Analysis of wireless communications was a strong point of mine. Leap Communications (brand: Cricket) was profitable as a long and short position. We were shareholders while it was expanding markets and enjoying the steep growth curve. It was a successful short position when the growth slowed and the churn and bad debt from the customer base significantly impacted profitability and growth. he stock eventually began to trade below asset value because of operational challenges and debt. Shortly after seeing Softbank acquire e-Mobile (#4 in market share) in Japan for the value of its network and spectrum despite a similarly challenged debt and growth profile, I looked at Leap through a different lens of the asset value. I turned the short to a long position and within a few months, AT&T acquired the company for a 100% premium to the share price. The lesson to be learned is not to get too focused on operational performance and not to lose sight of what a business' assets could be worth when combined with another business. Particularly if the assets are in short supply (spectrum) and the assets difficult to replicate (wireless networks).

Can you share a story of one of your worst investments and why it went so badly?

Classic mistakes on bad investments in my experience has been from overestimating founders / management teams. There is a big difference between founding and scaling a company and staying truly objective throughout. CEO's need to be extremely open to quality advisors and outside input.

There is a lot more focus now by investors, employees and customers on companies that do good (and do no harm) as well as make money. How do you think about this topic in the context of your investing?

Given how hard and complex it is to start, grow and compete as a company, I think you want to stay as focused as possible on the business plan. Achieving business goals clearly involves company culture and the ability to attract and retain talent. Therefore, the social objectives of a company should be appropriately tied to attracting and retaining talent. Beyond that, the company should be looking to maximize value created and let the shareholders take those returns on investment and reinvest into "doing good" as they see fit.

Who have been role models you most admire?

In investing, it was my peer Portfolio Managers at Citadel Global Equities San Francisco.

What do you believe are the most important traits of a technology entrepreneur?

Stay objective and stay paranoid.

What do you believe are the most important traits of an early stage technology investor?

Ask difficult questions and challenge management/founders when their product/service strays from the original pitch.

If you could tell a technology entrepreneur just one thing, what would it be?

Leverage your investors, their expertise and their networks to the benefit of growth in enterprise value.

What do you know now that you wish you had known earlier on?

Be more critical of management that makes excuses for failing to deliver on goals and milestones. Look for more personal accountability.

Would you please create a personal quote that captures important wisdom about participating in early stage investing?

"It's fine to be long term but even marathon runners watch their early pace. Investors need to establish their own milestones for the company to achieve."

All entrepreneurs must have tremendous tenacity. It is a long, hard journey with lots of rejection. Entrepreneurs need to believe in themselves but also build a network of advisors to lean on and receive advice from.

"You can be a huge benefactor just by making yourself available and trustworthy."

David Paul

Angel Investor

Keiretsu Forum, AngelList

What was your first career?

I had a thirty-year accounting and finance career in Silicon Valley manufacturing firms such as National Semiconductor, Intuitive Surgical and Fusion-io among others.

Why do you choose to focus on early stage technology investing?

I wanted access to the potential for outsize returns in a wildly speculative and defined portion of my portfolio.

When your investing is not going well, what do you do or say to yourself to keep motivated?

For this illiquid portfolio, I consider the money gone at payment. It's always going well, until it isn't. High tolerance of ambiguity is required.

What do you love most about what you do now?

I love the energy of the start-up world. The passion is palpable. It's not just another day in a big office.

What dreams do you have for the next 10 years?

Our three daughters are in their twenties and I expect that if we're going to be grandparents it will happen in that rough timeframe. My dream is to be able to spend time with those lovely creatures. If there is a homerun angel investment I would love to share the bounty of that with them.

If you could change one thing about the world we live in, what would it be?

I would eliminate unaccountable leadership.

What was your first technology investment and what happened?

I was intrigued by a software firm developing an enterprise resource management program for the small and medium business market. I had a good deal of context around why the market had room for an entrant like this. I had not learned the rigors of due diligence yet and jumped into a convertible note. It went well for a couple of years with an infusion from a VC that converted the note to shares. A cool stock certificate came in the mail. Things looked good and then the communication started to dwindle until finally "we're closing shop". It was about 4 years end to end. During that period through other collaborations and application to the process of continued education, I learned a great deal about what being an angel investor means and how to go about it.

Can you share a story of one of your best investments, and why it went so well?

My best investment involved a lot of conversation with the CEO of the company as the lead of our due diligence team. It was for an offering in the Internet of things (IoT) space with a software subscription overlay. The CEO led from his values that included a great deal of respect for all team members and humility around his own weaknesses. He was a guy that you could easily trust and support. He called on me occasionally to bounce ideas and get additional perspectives. He had a best in class quarterly communication letter for investors that laid out the good and the not so good. He had a positive exit in the first quarter of 2020 - fortuitous timing which I attribute to karma.

Who have been role models you most admire?

My boss at my first Silicon Valley job taught me what it meant to be a fiduciary and the high standard of ethics inherent in that responsibility. All you have is your reputation. He had my back at all times.

What is the best advice you ever received about how to succeed in Silicon Valley?

My business school management professor had a catch phrase that was part of his research *"Do what you say you will do."* (DWYSYWD) that I live by. Another colleague had a good thinker saying *"There's nobody smart enough to piss me off."*

What do you believe are the most important traits of a technology entrepreneur?

The best entrepreneurs are the ones who know what they do not know and are able to fill that gap through hiring or advisors. True competence is not pretending to know everything; it is borne of self-awareness. That humility and ability to actively listen allows one to take information in and effectively communicate a rationale for a decision and keep moving the organization forward. Trusting your team enough to delegate and accept results. They need to understand that their every move is being watched and if they don't have genuine care for the team it will get exposed.

What do you believe are the most important traits of an early stage technology investor?

Perspicacity and fortitude come to the fore. There are many unknowns for both the entrepreneur and the investor at an early stage. Asking good questions and keeping an open mind about the need to change course are valuable for the investor. Good entrepreneurs inevitably have to steer the company from time to time. Good investors will respond timely to requests for counsel so that the steering is not a surprise and is done with the benefit of input. A good investor will already have established this rapport before investing.

If you could tell a technology entrepreneur just one thing, what would it be?

Communicate, communicate, communicate. I guess that's three things. Oops.

What do you know now that you wish you had known earlier on?

It takes a minimum of twenty $25,000 investments to have a roughly 80% confidence of an average angel return. Forty investments grow that confidence level to over 90%. All investments should have 100 hours of methodical due diligence before investing which can be done through a team approach. That means that there will be a ton of hours spent that do not result in an investment. Have an investing thesis and stick to it.

Would you please create a personal quote that captures important wisdom about participating in early stage investing?

"Don't be afraid to engage and share the benefit of your experience with the entrepreneur when asked. That rapport should be established at the beginning. Competent ones have all the paranoia and doubts that you do and will welcome the help. You can be a huge benefactor just by making yourself available and trustworthy. Would you give money to someone you don't trust? It is not weakness to ask for advice, quite the contrary."

> *"Investing in early stage startups is investing in the world you want to live in."*

Victoria Pettibone

Venture Capital Partner and Angel Investor

Astia, Astia Fund and Astia Angels

What did you want to be as a child?

An astronomer, except I kept telling people "astrologer" because I thought that was the correct term, and I never understood why people laughed at my desire to study outer space!

What was your first career?

Casting Director for the Broadway show *Rent*.

What other careers have you had before your current one?

Entrepreneurial – I co-founded and ran a social enterprise in the media space for eleven years, developing and producing theater and film written by women and running a media literacy program for teenage girls.

Why do you choose to focus on early stage technology investing?

I focus on early stage investing (all high growth sectors) because I love getting inspiring ideas off the ground, and working directly with visionary entrepreneurs.

When your investing is not going well, what do you do or say to yourself to keep motivated?

You have to know that some of your investments will fail so it's important to invest in projects and entrepreneurs you deeply believe in. When an investment is tanking, I try to learn from the experience – are there any lessons that I can apply to my next investment, what could I have done differently?

What do you love most about what you do now?

As a producer in my former career, I loved opening night when an entire audience would be captivated and experience something magical and I knew that I had made that happen. It is the same with investing – of course it's the entrepreneur doing all the hard work, but without the investors providing that initial funding, the project would not be out there, making an impact, creating an experience for the rest of the world.

What dreams do you have for the next 10 years?

When I started my career, I created a vision plan that included the legacy I wanted to leave past my death. I have veered off that initial path tremendously (it did not involve venture at all!). Changing careers, which I did in my mid-thirties, is a scary, unnerving thing to do. I felt unmoored, unsure of the next right step. I thought microfinance was the road not taken, as I had spent time in Sub-Saharan Africa in high school and college studying early days of microfinance, and that led me to graduate school, but then I started learning about the dearth of venture funding to women entrepreneurs. It was a problem I felt compelled to address. As I began to work in that field, I had my first child and three years later my second. I call that time period "the exhaustion years" – I was working and extremely productive, but I was so drained emotionally, psychologically, physically, the idea of envisioning ten years down the road let alone 1 year was beyond me- all that mattered was to keep these little children alive and get my work done! I'm fairly recently out of the exhaustion years (yes working moms of very young children, you do get through them!), and I've found myself starting to imagine longer term again, but beyond dreaming about having wildly successful fund performance and seeing the investment landscape include women and entrepreneurs from other underrepresented groups, I've not done a new, robust visioning exercise quite yet. I don't want to force it, I want it to be inspired, creative, fun and exciting. In short, check back with me in a year!

If you could change one thing about the world we live in, what would it be?

Right now the most important thing is that we get responsible, smart, thoughtful, empathetic, courageous leaders into positions of power – from the top of government to the local.

Why do you think the innovation economy/ Silicon Valley has a poor track record on diversity and inclusion? What needs to change?

I think Silicon Valley has a poor track record on diversity and inclusion because the traditional venture model is broken. It relies on networks and on being an "insider." Decisions are driven by an overreliance on the notion of *trust,* which is inherently biased, and on the misconstrued notion that if something seems familiar it is more likely to succeed.

Early stage investing is certainly a gamble – no one really knows what is going to succeed, who is going to execute well, what factors out of their control will have an impact the startup can't withstand…the companies are too early to have enough historical data to rely on. Early stage investing is a leap of faith, which is where this problematic notion of "trust" comes in. VCs are more likely to "trust", and therefore invest in, entrepreneurs that are already in their network or that talk like them, dress like them, went to their school, know the same people they do etc.

When the predominant phenotype of the VC is what it is today, and deals are sourced through a predominantly white, male network, we miss tremendous opportunities. What needs to change is a realization that presumptions around "trust" are totally wrong and racially, geographically, and gender biased. The venture industry needs to recognize that profitable opportunities come from diverse teams, and that diverse teams outperform (there is significant data demonstrating this). With this reckoning, we can work to level the investment playing field, and in turn increase the diversity in Silicon Valley companies. And finally, we need to look at increased gender and racial diversity among venture partners.

There is a lot more focus now by investors, employees and customers on companies that do good (and do no harm) as well as make money. How do you think about this topic in the context of your investing?

Great investments are those that solve huge problems. We live in a time with vast societal, environmental and economic problems that desperately need solving. Companies that provide novel, efficient, better solutions to these problems stand to be enormously profitable, and therefore, good investments. In terms of my own investing, because I believe that inclusive teams make better decisions, I invest only in companies that include women and I place a premium on a team with additional layers of diversity, especially around race.

Who have been role models you most admire?

The CEO of Astia, Sharon Vosmek, is a visionary leader and a terrific role model. She is a thought leader in the space of women in the economy and a savvy investor in the startup ecosystem.

What do you believe are the most important traits of a technology entrepreneur?

All entrepreneurs must have tremendous tenacity. It is a long, hard journey with lots of rejection. Entrepreneurs need to believe in themselves but also build a network of advisors to lean on and receive advice from.

What do you believe are the most important traits of an early stage technology investor?

Curiosity, good listening, perseverance. Curiosity drives an investor to deeply learn about a company, good listening enables them to understand what the entrepreneur needs and how they can best support them, perseverance enables them to stay calm for the long haul, because most investments take much longer than expected to exit.

Would you please create a personal quote that captures important wisdom about participating in early stage investing?

"Investing in early stage startups is investing in the world you want to live in."

It's a culture of outsiders, upstarts, and troublemakers. This trickles down to everything, from the labor laws to the coffee people drink. Everything deserves to get questioned. Everything yields to the innovator, to the creator, to the mad scientist who has a vision about how the world ought to change.

"Focus on areas in which you're comfortable and knowledgeable and spread your wealth and time according to your passions."

Mary Jo Potter

Angel Investor

Keiretsu Forum & Capital, Portfolia, Healthcare Angels, Purple Arch, The Band of Angels, Skydeck

What did you want to be as a child?
Nurse.

What was your first career?
Consultant to high tech around sales and management.

What other careers have you had before your current one?
Teacher, Retail Exec, entrepreneur. Mainly growing companies in an executive role

Why do you choose to focus on early stage technology investing?
I felt it was the best way for me to help change the healthcare industry.

When your investing is not going well, what do you do or say to yourself to keep motivated?
Leave it for a day, and come back with a purpose. Also, I have to reframe often from short term to long term.

What do you love most about what you do now?
Buoying up the entrepreneurs and their teams. Meeting and engaging with very sharp people.

What dreams do you have for the next 10 years?
Having 5 of my invested companies impact over one million people in their health and wellness.

If you could change one thing about the world we live in, what would it be?
Having more capacity to care about those beyond ourselves.

What was your first technology investment and what happened?
Apple in 1980's. I actually learned a little coding from my father, who was always an early adopter. I discovered that was not my long term calling! Then I invested in Supercuts, and they paid out a dividend for 25 years.

Can you share a story of one of your best investments, and why it went so well?
I invested in Les Concierges and later became a board member. We sold the company twice and it is now part of Accor. It was a challenging ten years, but worked out well for all.

Can you share a story of one of your worst investments and why it went so badly?
When we invested in Mentura it was a Netflix equivalent for family friendly content. Things went well for about 18 months, and then all of a sudden the founder disappeared. It ended up that he was defrauding the company. We fought for a return of capital for two years, and ended up with 30 cents on the dollar, but at least the founder won't be starting another company anytime soon.

In your opinion, what is so special about Silicon Valley?
It is a holistic environment that accommodates innovation in a powerful way. Having all the elements of support be so proximate (money, thought leadership, universities, previously successful entrepreneurs, flexible legal and accounting talent, and incubators and accelerators that are world renowned), encourages startups to dive in, and when they run in to trouble, there is a community of empathy to work with them.

Why do you think the innovation economy/ Silicon Valley has a poor track record on diversity and inclusion? What needs to change?
Silicon Valley has been very successful with immigrants creating and establishing start-ups. They have not done as well with women and minorities. This is a two-headed monster. Many women and minorities have to overcome conscious and unconscious bias, but they are not always willing or able to take the level of risk needed to be successful as entrepreneurs. The best way to further

these populations is to have representatives of those communities be much more visible and involved in everything from funding to development efforts and board positions.

There is a lot more focus now by investors, employees and customers on companies that do good (and do no harm) as well as make money. How do you think about this topic in the context of your investing?

I think this is about the preference of many workers to be attached to a higher purpose than just making money. This is especially true for millennials. This phenomenon is also playing out in the branding world. This affects investors in that they look for passion and purpose most especially with the founders. Most aren't willing to lower expectations of returns, but all things being equal, like having mission and money purposes integrated.

Who have been role models you most admire?

My parents, Jim Woolwine (founder of Presidio Bank), Tom Bertelson (a fellow board member at Hanna), Anne DeGhest (investor on healthcare) and Condaleeza Rice (leader in government and education).

What is the best advice you ever received about how to succeed in Silicon Valley?

Keep on keeping on. Most give up too soon, and might only be one opportunity away from a major inflection point.

What do you believe are the most important traits of a technology entrepreneur?

Persistent, risk taker, flexible, high integrity, willing to learn, likeable, and high energy.

What do you believe are the most important traits of an early stage technology investor?

Optimist, patient, skeptical, well-connected, collaborative, able to lose money, diversified, open-minded.

If you could tell a technology entrepreneur just one thing, what would it be?

Make sure what you are doing is really a must have not a nice to have.

What do you know now that you wish you had known earlier on?

As an investor, it is better to have 10 times $10,000 investments than one $100,000 investment.

Would you please create a personal quote that captures important wisdom about participating in early stage investing?

"Focus on areas in which you're comfortable and knowledgeable and spread your wealth and time according to your passions."

Leverage the vast resources of information, experience and expertise that are available here. The Valley is a very open platform and people are willing to share and support you if they believe you're bringing value to the table.

> *"You have to make a bet on something that no one believes in yet, and that requires peering into the future and having conviction about how the world is going to change."*

Hasseb Qureshi

Venture Capital Partner

Dragonfly Capital

What did you want to be as a child?
Author / astronaut / basketball player.

What was your first career?
Professional poker player.

What other careers have you had before your current one?
Professional poker player, software engineer.

Why do you choose to focus on early stage technology investing?
Entrepreneurs are my people. I'm cut from the same cloth. I love nurturing and encouraging people to take risks and create new things.

When your investing is not going well, what do you do or say to yourself to keep motivated?
Innovation and openness always wins in the long run. See the Internet—it took decades to reach fruition, but it looked hopeless many times along the way.

What do you love most about what you do now?
Blockchain is completely reinventing finance and political structures. Keeping up with this space forces you to constantly learn new things and challenge your old assumptions about the world.

What dreams do you have for the next 10 years?
I'd love to build a generation-defining blockchain investment firm. I want people to look back and agree that this firm changed the way that blockchain investing is done, the way that Sequoia or Andreessen did as technology investors.

If you could change one thing about the world we live in, what would it be?
Very simply: that there would be more of a positive-sum thinking in the world—consider technology, trade, politics. It seems that zero-sum thinking dominates headlines and punditry these days. That worries me for the future.

In your opinion, what is so special about Silicon Valley?
It's a culture of outsiders, upstarts, and troublemakers. This trickles down to everything, from the labor laws to the coffee people drink. Everything deserves to get questioned. Everything yields to the innovator, to the creator, to the mad scientist who has a vision about how the world ought to change.

In the culture of the upstart, failure is not just accepted, it's celebrated. YCombinator has institutionalized this—it's impressive to have been a YCombinator company, even if you folded! It's easy to forget what an incredible cultural accomplishment this is. In most places, having a failed startup is a pockmark on your career. Silicon Valley lionizes its failures. There is unforgettable street cred that is reserved only for those who have lived to tell the tale. It's incredibly hard to build that culture twice. That's why every city that has tried to copy Silicon Valley has failed.

There is a lot more focus now by investors, employees and customers on companies that do good (and do no harm) as well as make money. How do you think about this topic in the context of your investing?
I'm of two minds on this. First, it's clear that tech companies have become incredibly powerful in our society, and as such they have come to take on dramatic influence in the provision of public goods (think Google, Facebook, Apple, etc.). We have to hold them to high standards. At same time, it's also clear that we don't want to be in the situation where we are relying on private actors and companies to enforce our social norms. Holding them to the same standards that we'd hold a democratically elected government betrays a failure of the government itself to shape a just society.

So given all that, you have to ask: what can you actually practically do as an investor? If you'd backed

Google or Uber in the early days, you couldn't have foreseen the effect they'd eventually have on the world. The world is complex and unpredictable, and every story of a company affecting something negatively must be weighed against the value that innovation inevitably creates, which is always harder to measure or point to in the short run. (In the long run, it's most of what has improved the human condition over history.) The most we can ask of ourselves is to be mindful of tradeoffs. Not to invest in things that you don't believe are truly innovative. And finally, to remind your entrepreneurs of what should be their northstar: creating true value for people.

What is the best advice you ever received about how to succeed in Silicon Valley?

If you want to be successful, get really good at something that the world doesn't already know how to do. If other people know how to do it, the world can train people at the ready. To be really valuable, you must blaze a trail that no one knows how to walk yet.

What do you believe are the most important traits of a technology entrepreneur?

Vigorous curiosity, an abnormal disregard for "the way things are done," and a bent toward action and experimentation.

If you could tell a technology entrepreneur just one thing, what would it be?

The road is longer than you think. That's true even if you take into account that the road is longer than you think.

Would you please create a personal quote that captures important wisdom about participating in early stage investing?

"Every form of investing is about buying valuable things for less than they're worth. To that end, early stage technology investing is ultimately about being non-consensus right. If everyone already believes in X (whether X is a technology or an entrepreneur), then the price will probably be right. That's not how you're going to find success. You have to make a bet on something that no one believes in yet, and that requires peering into the future and having conviction about how the world is going to change (and who is going to change it)."

> *"Early stage is the riskiest but perhaps the most rewarding phase of the investment process."*

Sean Randolph

Think Tank Leader

Bay Area Council Economic Institute

What did you want to be as a child?

First an archaeologist, later an architect.

What was your first career?

My first career was in the Federal government, focused on international affairs: on House of Representatives staff, White House staff, and in senior positions at the Departments of State and Energy.

What other careers have you had before your current one?

For a number of years I ran an international organization of approximately 1000 major companies in Asia and Pacific Latin America. With some consulting in between (still focused on Asia and Pacific Latin America) I later served as the international trade director for the State of California. In contrast to previous roles working for the state had nothing to do with policy. It was all about helping California companies, mostly small and medium sized, to develop overseas markets.

Why do you choose to focus on early stage technology investing?

This is where the most creative, impactful, and innovative companies will come from. Some will become great companies that will transform economies and how we live.

When your investing is not going well, what do you do or say to yourself to keep motivated?

As an analyst/advisor I see trends. Risk comes with the territory. For investors, believing in a company's ultimate potential and the founders' vision is essential.

What do you love most about what you do now?

I get to see the big picture and how the pieces fit together, in Silicon Valley and around the world. Sharing that perspective is incredibly motivating.

What dreams do you have for the next 10 years?

I want to help people understand Silicon Valley's role in the innovation process and how it's evolving, but also help Silicon Valley see how the global environment of innovation is evolving. There is so much talent and creativity around the world, between technologists and highly motivated founders, that the potential to improve our economies and our quality of life is enormous. I want to be in the middle of that process, helping it come together.

If you could change one thing about the world we live in, what would it be?

I would replace a zero-sum perspective with an appreciation of the potential that is available through collaboration, particularly across borders. That includes a deeper recognition of the value of global talent and of the fluid movement of data and ideas. At this moment, we seem to be at some risk of losing that. On the flip side, we also need to build bridges to draw out the entrepreneurial talent and creativity that's available across the US, not just in Silicon Valley.

What was your first technology investment and what happened?

Our regional (Bay Area) investment fund made several early calls. One was Brightsource. We saw renewable energy at utility scale as being an area with large potential, particularly given the technology being produced in the Bay Area and the scale of the market for renewable energy being supported by forward-looking policies in the State if California. We also believed that a transition to renewable energy was an essential societal goal that would ultimately lead to the creation of great companies. Brightsource subsequently developed large solar thermal generating facilities in Southern California and later entered global markets. Our participation was small, but its focus proved correct.

Can you share a story of one of your best investments, and why it went so well?

Another of the early investments we were directly associated with was Tesla. At the time we had a "double bottom line" fund that looked not just at the potential for returns but also at the potential for a positive economic impact in the region where we're located – the San Francisco Bay Area. That obviously worked out well, as Tesla has become profitable and is arguably leading the global EV market. It also delivered results for the regional economy, as Tesla is now the largest manufacturer in the Bay Area and anchors an extended supply chain of automotive and energy suppliers in the region.

In your opinion, what is so special about Silicon Valley?

What makes Silicon Valley so special is its unique combination of assets, both tangible and intangible. That includes some of the best universities in the world and the technologies and IP they produce, the world's largest concentration of private risk capital, successful entrepreneurs who have become investors and mentors, a perspective that embraces risk and seeks and rewards transformational change, and open environment that attracts talent and is deeply connected to the global economy. It's not the only place in the world where you can find those assets, but they don't exist anywhere else with the same intensity or to the same degree.

Why do you think the innovation economy/ Silicon Valley has a poor track record on diversity and inclusion? What needs to change?

This really goes back to Silicon Valley's first generation of founders: most were men who were engineers or technologists from universities or from large companies or who came to Silicon Valley from overseas. And most were intensely focused on the survival and growth of their companies and not on other objectives. And that's where most Venture Capitalists also came from. So you have to look at the pool. At the same time, not many women (comparatively) were pursuing careers in engineering or computer science.

It takes time to change those patterns. In the near term the Valley needs to more consciously support and promote women and minority executives. In the longer term, more women and minorities need to make their way through the higher education system and into the economy in order to connect with those opportunities.

Despite the Valley's obvious shortcomings in this area, I think it's also important to point out that the Valley is incredibly diverse in terms of global participation: a high percentage of founders comes from overseas, particularly from China and India.

Engineers, founders and investors are also here from in Europe, Mexico and Southeast Asia – everywhere. From that perspective it's a very open environment. That's another form of diversity where we've done better.

There is a lot more focus now by investors, employees and customers on companies that do good (and do no harm) as well as make money. How do you think about this topic in the context of your investing?

Yes, this is a growing focus. Investing will always be drive by ROI. If there's no ROI there's no opportunity to d good, and the more successful the company the larger that opportunity becomes. So there's no getting away from that. But investors are coming in at the earlier stages of the companies they invest in, and particularly where there's an opportunity for large scale or transformational change their investment can make a difference. Having that perspective as part of the investor's core goals or values is important. In Silicon Valley many employees of technology and other companies look for that as well, which plays into the national debate on how the benefits of growth can be more widely spread. Our own experience – investing in renewable energy and in electric vehicles – confirms that doing well and doing good are very compatible goals.

Who have been role models you most admire?

My role models come from the political world. Ronald Reagan, who I worked for, showed me the importance of having a clear idea and communicating it well (leadership). Secretary of State George Schultz, who I also worked for, taught me the importance of understanding interests and how to reconcile them.

What is the best advice you ever received about how to succeed in Silicon Valley?

Leverage the vast resources of information, experience and expertise that are available here. The Valley is a very open platform and people are willing to share and support you if they believe you're bringing value to the table.

What do you believe are the most important traits of a technology entrepreneur?

Investors look for founders and leadership teams they believe can deliver. A good technology entrepreneur needs to be able to distinguish their company from the pack by succinctly communicating his or her vision, recognize critical skills gaps and fill them with other motivated people, motivate their team, live with risk and uncertainty, and persevere in the face of existential challenges.

What do you believe are the most important traits of an early stage technology investor?

Recognize where markets are going or where new markets will be created (identify needs that aren't being met) and recognize talent (founding teams that can deliver).

If you could tell a technology entrepreneur just one thing, what would it be?

These days in particular I'd say pivot to the market. COVID has changed the startup environment, undercutting some sectors, increasing interest in others, and accelerating digitalization across the board. There's plenty of capital waiting to be invested if you can fill the right need. And when you get the investment watch the bottom line: in times of uncertainty managing costs, creating revenue, and establishing a path to revenue and profitably has grown in importance.

Would you please create a personal quote that captures important wisdom about participating in early stage investing?

"Early stage is the riskiest but perhaps the most rewarding phase of the investment process, as promising entrepreneurs are identified before they become well known, and where the potential to catalyze growth opportunities is greatest. Alignment with other investors who have similar motivation and vision and who are prepared to share that risk and opportunity is critical to success."

> *"Invest if you believe that the problem is big, the solution is inevitable and your team is the best team to make it happen."*

John Ricci

Angel Investors

Managing Director US Angels, Keiretsu Forum, HBS Alumni Angels

What did you want to be as a child?
Engineer.

What was your first career?
Engineer.

What other careers have you had before your current one?
Executive in large corporations, Management Consultant, Entrepreneur.

Why do you choose to focus on early stage technology investing?
Because early stage technology investing is the bleeding edge where you see entrepreneurship and entrepreneur's dreams in their most original, rawest, unformed sate.

When your investing is not going well, what do you do or say to yourself to keep motivated?
You need to kiss a lot of frogs and sometimes you find the princess. If the Gods are smiling, she's riding a unicorn.

What do you love most about what you do now?
Talking to inspiring, interesting, sometimes crazy, never boring people who want to solve problems and change the status quo, sometimes change the world.

What dreams do you have for the next 10 years?
I want to:

a) Contribute bring their innovation to the US and raise financing globally. The first step in that journey is building a growth accelerator and financing platform, under the guise of several mentor driven programs offered to later stage foreign startups;

b) Do one last startup, maybe in the area of age-fighting neutraceuticals and other non-FDA-regulated products.

c) Get in better shape, climb a mountain and do a 100K bike ride

d) Start a non-profit with a 300 year vision that every politician and rule maker on earth will sign a gender equality pledge.

What was your first technology investment and what happened?
My first technology investment was in a company started by two classmates from business school. They had this really cool device that basically functioned as a computer mouse on your head. Imagine that: you could point at any part of the screen by simply moving your head which naturally followed the direction of your eyes; you could even click, using another part of the device that you would press on with your foot, and your hands would stay on the keyboard and just keep on typing. I thought this was the best thing since sliced bread, and everyone would want to free up their hands from that horrible device called a mouse and purchase one. Needless to say, that did not turn out to be the case, an even the folks who could really use it such as people with disabilities turned out to be a very limited market with limited purchasing power. The company and its technology ended up being acquired for a small price, and my dreams of a large payout never came true. I did learn a great lesson though. Before investing in a company, make sure the "dog wants to eat the dog food". I look at over a thousand startup companies every year, and I am still amazed by how often they fail because their potential customers just do not really need or want their products or they do not want them badly enough.

Can you share a story of one of your best investments, and why it went so well?
My best investment was not in a technology company. It was in what turned out to be one of the category leaders in craft beers. One of my colleagues in the consulting company where I worked approached a bunch of us and said:" I want to quit my job and start a beer company that will sell European-style, great tasting beer to bars and venues in Boston and then all over the East Coast. I have a recipe from my Austrian grandfather and I have found unused brewing capacity. I am going around asking everyone to invest. I think people are really thirsting for a better beer and I have gone to the bartenders in the

Boston Back Bay and they are willing to give it a try. I am hiring away my secretary who is going to be my first salesperson and she will have shares. Everyone is willing to talk to her. Come on to my house, taste the beer and bring your checkbook." Since I had grown up in Europe, in an area that was actually renowned for its regional beers, the story made sense so I wrote a small check. The company did extremely well. As the saying goes, the "stars were aligned" for this one. This product was excellent and highly differentiated. The timing was perfect for a pioneering, category-defining craft beer; and the team was great: smart, passionate, driven, relentless and agile.

Can you share a story of one of your worst investments and why it went so badly?

One my worst investments happened because I fell in love with an idea, a business model and I became blinded to the shortcomings and the lack of experience of the management team. The company could have turned out to be an early home-run in the difficult category but large category of digital marketing solutions for main street merchants and small businesses. At the time, Yelp was the dominant solution, and most merchants hated them. No one had really managed to replace the Yellow Pages and local Google search and Facebook advertising were not yet easy-to-use. The business model involved the clever use of a 100% commission, low-end salesforce, which would be economically viable but extremely challenging to manage. The founders raised more than a million dollars but the CEO, even though this was his first startup, proved to be difficult to coach. Instead of slowly refining his execution in a couple of test markets, he instead adopted a land grab strategy and started in a dozen markets at once. Mistakes were amplified and the company soon ran out of money. He was let go, but the right replacement could not be found before the company was forced to shut down. The moral of the story is obvious to any investor who has had some failures: there is nothing more important that the team, and make sure the founders have both the courage to defend their own ideas but also the wisdom to be coached and listen to others' input.

In your opinion, what is so special about Silicon Valley?

Many of the reasons Silicon Valley is so successful are well documented. Still here are what I believe are the two most important factors that help make Silicon Valley so special and so difficult to replicate: First is the extent to which Silicon Valley has an immigrant, risk-taking culture. Immigrants have always been, out of necessity, people willing to take risks, work very hard, disregard privileges and tolerate failure. The United States is a country born from immigration. But once in the country, where did our forefathers go for more opportunity. The answer is California. In a sense the Gold Rush mentality has defined Silicon Valley -and Hollywood- ever since the first miners and the first movie producers. To this date, Silicon Valley is home to one of the largest foreign-born populations in the US (38% of the nearly 2 million people living in the San Jose-Santa Clara metro area are foreign-born). And many of these foreign born residents are skilled, educated, entrepreneurial and willing to take risks and start companies. Second is the *density* of the ecosystem. In a fairly limited geographical area -4 counties if one adopts a somewhat looser definition of Silicon Valley that includes San Francisco and part of the East Bay-, there are more startups, more engineers, more innovation, more great universities, more coding students and biotech PhDs who want to start new companies, more mentors, more meet-ups, more startup lawyers, and last but not least more venture capital money than anywhere else in the world.

Why do you think the innovation economy/ Silicon Valley has a poor track record on diversity and inclusion? What needs to change?

The innovation economy is based on STEM education. Our society is failing women and minorities on two levels. First women are not systematically encouraged to explore and thrive and have fun in scientific matters in school, starting at a young age and in a way that systematically fights the "girls don't do math or coding" syndrome. Minorities, especially "underperforming" minorities from lower income families, are in even worse shape because the education and societal structures condemn then to failing and underfunded schools, poor learning environments and ill adapted peer and role models. Second, Silicon Valley culture, with its egalitarian streak, and its short term "you have the skills or you don't" focus, is unable or rather unwilling to put in the considerable effort it would take to select, support and train as needed women and non-Asian minorities. What needs to change at a society level is the political will to give more funding to minorities and STEM training and change curricula. That would require a significant political push, almost a Marshall plan, which may be difficult to achieve in the short term. Meanwhile and in parallel, Silicon Valley should invest an order of magnitude more effort and money into correcting the defaults of the current education and societal systems by introducing much more aggressive and systematic diversity and inclusion goals, maybe even self-imposed quotas.

There is a lot more focus now by investors, employees and customers on companies that do good (and do no harm) as well as make money. How do you think about this topic in the context of your investing?

Very simply, I do not invest in anything that does harm. I do not necessarily focus on socially responsible investments or double- or triple-bottom-line investment as I believe those take a different investor profile and a different ecosystem. I am still mostly focused on profit and commercial success as the main yardsticks. But I also believe that current market trends will reward companies and their investors when these companies' products align with consumers' increased interest in healthier and more sustainable, "goof for society" solutions.

Who have been role models you most admire?

In the field of business, Steve Jobs for his relentless focus on "making" things that consumers love using, Elon Musk for thinking outside the box and Clayton Christensen, a former colleague of mine, for sheer analytical brilliance. Clayton is the guy who taught me about the hidden costs of complexity. In the field of politics and despite what one may think about his achievements, Barak Obama for taking on and navigating the minefields of the first black presidency with grace, elegance and humor. In the narrower field of angel investing, Randy Williams, the founder of Keiretsu Forum, for bucking the trend and building the largest angel investor ecosystem in the world, while charging companies that were trying to raise funds to compensate for the efforts required in running angel investor groups, a model that everyone else said would never work.

What is the best advice you ever received about how to succeed in Silicon Valley?

Love what you do and stay relevant. At the same time adapt, adapt and adapt some more. Silicon Valley can be a difficult place to navigate as people age and they may be required to learn new skillsets. Stagnation and complacency are the enemies of success in Silicon Valley, both for people and companies. Innovation is always creating a new and different tomorrow. You've got to stay and love staying on your toes.

What do you believe are the most important traits of a technology entrepreneur?

A passion for building great products, the wisdom to be coachable, extreme resilience and agility, the willingness to make tough decisions quickly and first and foremost the ability to attract and work with great teams.

What do you believe are the most important traits of an early stage technology investor?

A great investor often:

1) Has the ability to discover great companies;

2) Takes a disciplined investment philosophy with a portfolio approach, where the investor invests in companies he understands or alongside others who understand them, according to a well-designed set of criteria;

3) Has the time to do the required due diligence "homework: or can leverage other people's work;

4) Has the ability to read and assess people along a variety of dimensions;

5) Invests both financial and human capital and is wise enough to know and ask whether founders need and want the help they may be offered.

If you could tell a technology entrepreneur just one thing, what would it be?

Make sure the world is craving for what you want to sell.

What do you know now that you wish you had known earlier on?

I wish I had known earlier that good angel investing, if one does it for the returns, is a game of chasing home runs. I would have avoided some of my early investments that ended up doing OK, but were not rewarding enough to compensate for the ones that did not work out. I also wish I had understood earlier that angel investing is somewhat comparable to fishing. Some people go fishing to catch big fish; others just enjoy the fishing. Of course no one wants to return empty handed, but you need to know what kind of fisherman you are.

Would you please create a personal quote that captures important wisdom about participating in early stage investing?

"Invest if you believe that the problem is big (the horse race has a big prize), the solution is inevitable (you have the best horse) and your team is the best team to make it happen (you have the best jockey)."

> *"As an entrepreneur, there are two questions that are the most important to answer: Are these the problems I want to solve? And, are these the people I want to solve them with?"*

Jana Rich

Executive recruiter and diversity advocate

Founder and CEO, Rich Talent Group

What did you want to be as a child?
As a child, I wanted to be a singer! Even though I'm not really qualified to be one...

What was your first career?
I started my career in PR, thinking I was on the path to be a CMO one day.

What other careers have you had before your current one?
Before I became an executive recruiter, I worked in public relations and management consulting.

Why do you choose to focus on early stage technology?
As an entrepreneur, I love working with other founders to build their teams. And I believe younger high-growth companies (as well as established ones), need a partner that is exclusively focused on diversity and inclusion—especially at the senior-most level, which is where I focus.

When your work is not going well, what do you do or say to yourself to keep motivated?
I stay focused on our mission, which is diversifying teams. I also get inspired by our portfolio of companies—there's always something new and exciting happening.

What do you love most about what you do now?
What I love most about what I do is working with a team and founders who share our company's mission: to build diverse, transformative teams.

What dreams do you have for the next 10 years?
My hope for the next 10 years is that teams become stronger because they have more voices from women, people of color, and the LGBTQ community.

If you could change one thing about the world we live in, what would it be?
There would be true equality regardless of your race, gender, or sexual orientation.

In your opinion, what is so special about Silicon Valley?
What's so uniquely special about Silicon Valley is the incredible support for entrepreneurship. It's actually ok, and even encouraged, to fail. You can pick yourself back up again and still create something viable or fundable. And though there's still a lot of work to be done when it comes to diversity, I do feel there's a respect for it here, compared to some other parts of the country.

Why do you think the innovation economy/ Silicon Valley has a poor track record on diversity and inclusion? What needs to change?
One of the main challenges to diversity in Silicon Valley starts with the belief that technology (as a function) rules everything. And technology tends to be the area with the least amount of diversity. So while it's important to get more diverse students into STEM majors, that shouldn't be the only emphasis. Companies have to focus on giving equal weight to other voices around the leadership table. Marketing and People functions, for example, typically have more diversity, so it's important to make space for their expertise as well.

Who have been role models you most admire?
I really admire Mellody Hobson, President and Co-CEO of Ariel Investments, for her courage to speak strongly for what she believes in and her passion for being a voice for women and people of color.

What is the best advice you ever received about how to succeed in Silicon Valley?
When John Doerr from Kleiner Perkins came to speak to my class at Stanford Graduate School of Business, I waited in a long line to meet him. I respected him so much—his portfolio had the kind of companies I wanted to work for. But I felt insecure about going into recruiting. Would that be impactful? When I got my chance to talk to him, he told me that even as a venture capitalist, he spent 50% of his time recruiting. Recruiting is incredibly important to a

company's success, he said, and I took that to heart.

What do you believe are the most important traits of a technology entrepreneur?

Tech entrepreneurs have to have a real mission and calling for the product or service they're creating, and also for building their team. That's the winning combination.

If you could tell a technology entrepreneur just one thing, what would it be?

Build a really strong team early. Surround yourself with people who complement you and are different than you.

What do you know now that you wish you had known earlier on?

I spent 25 years working on diversity and inclusion, and it felt like the world didn't pay much attention. I wish I knew earlier how important it was going to become, because it can be hard to stay focused when no one else seems to notice.

Would you please create a personal quote that captures important wisdom about participating in early stage technology?

"As an entrepreneur, there are two questions that are the most important to answer: Are these the problems I want to solve? And, are these the people I want to solve them with?"

"The only certainty in early-stage investing is that plans change and the unexpected often happens."

Daniel Rosen

Venture Capital Partner

Commerce Ventures, Board member of Blooom, ClickSWITCH, Kin, Socure

What did you want to be as a child?
The first profession I remember admiring was architect. The role fascinated me because it involves designing something from scratch, and those designs becoming realities in our world as the places people live, work and celebrate together.

What was your first career?
In 1999, my first job out of undergrad was as a business analyst for a large systems integration and consulting firm called American Management Systems. I was in the division that helped design and implement lending systems for large banks. It was less than a year before I found myself drawn into the world of startups and venture capital, however, and I've been in this world ever since.

What other careers have you had before your current one?
As a teen, I had a variety of less glamorous jobs – although I'm not sure I'd call those careers. My jobs ranged from paperboy to drugstore clerk to IT help desk.

Why do you choose to focus on early stage technology investing?
I am addicted to my role of supporting bold entrepreneurs as they create something from nothing and help transform the world in the process.

When your investing is not going well, what do you do or say to yourself to keep motivated?
VC investing is about great decades, not good or bad days. When having a bad day, I remind myself that tomorrow will probably be better.

What do you love most about what you do now?
If I do my job right, I can positively impact the journey of entrepreneurs who are trying to do the impossible: make something from nothing.

What dreams do you have for the next 10 years?
My dream for the next 10 years is to see my firm, Commerce Ventures, become a lasting institution and watch my team grow and succeed along the way. I hope to see dozens of our entrepreneurs achieve exciting commercial and financial success, while fundamentally improving the industries in which they and we operate. Finally, I hope to deliver to our investors the fantastic financial rewards they deserve (along with my heartfelt gratitude) for investing in us from the early days.

If you could change one thing about the world we live in, what would it be?
I wish our world were a fairer world, where opportunities were made equally available to people from all walks of life. Unfortunately, I see every day that people and companies benefit from so many, unevenly shared advantages. As a believer in the transformative power of startups on industries and careers, I feel it is our responsibility to help expand access and opportunity to all those with the will to achieve greatness.

What was your first technology investment and what happened?
I'm reminded of the first company I saw from initial investment through to successful exit. We invested in a seed stage mobile advertising startup in late 2006, with which I worked closely for the next three years. Along the way, I got to see almost every aspect of the VC-entrepreneur relationship from first term sheet to follow on financings, mundane board discussions to strategic off-sites. It was fascinating a talented group of early entrepreneurs and early employees literally help shape a new industry and then navigate the ups and downs, ins and outs of competing and positioning for success all along the way. In late 2009, the company was acquired for a considerable sum by one of the world's largest tech companies. When you see these stories all the way through, you realize that great successes are most often achieve by great teams…but often assisted by great timing. It's hard to win without both.

Can you share a story of one of your best investments, and why it went so well?

One of my most successful investments is in a provider of software and payment processing to small and medium-sized businesses. I had tracked the company from its early days, but it was not a fit for my prior firm's investing strategy. Venture capital is a relationship business, and I built and maintained a relationship for many years after my first meeting with the company, because I admired the company's CEO and believed I would continue to learn and grow by knowing him. Not long after I started my own firm, this CEO mentioned to me that he was raising a new round of funding and asked if we might be interested in participating. While the company at that time was more mature than our typical investment profile, I had developed such strong conviction about the CEO and his business over the years, that I believed it was worth the making an exception in terms of stage fit.

Since our investment seven years ago, the company has grown its revenues over 20x, completed a very successful IPO and is worth approximately $7 billion as of June 30, 2020. The drivers of their success included 1) strong founder-market fit (the CEO really understood the problem he was solving), 2) solving an important problem (automating payables), 3) delivering an easy to use interface (critical for SMBs), 4) utilizing strong channels to market (accountants and banks) and committing to SaaS before it was popular (SaaS architecture made distribution and adoption much easier).

Can you share a story of one of your worst investments and why it went so badly?

My worst investment was in a startup that we would today consider a 'neobank'. The CEO had a highly relevant background and the business plan was very exciting, but the company was too far ahead of the market and, in the end, that led to its failure and a complete write-off of our investment. In this case, being too early meant that the company had to build on inflexible, legacy infrastructure that prevented the company from being nimble in terms of new products and market tests. They were early in the proliferation of social media advertising and thus customer acquisition was also much more expensive than it would be in today's market. Access to financial data was also fundamentally more difficult at that time, which meant that opening and funding accounts involved much more friction than it would today with

modern data aggregation capabilities. Finally, there were very few sector-focused investors interested in FinTech at that time, which made fundraising quite challenging. Inevitably, the company ran out of cash because they couldn't find investors who shared the company's vision of an all-digital primary financial institution. I have learned a lot of lessons from that experience, but the main one is that being too early to a market opportunity can often be just as bad as being too late.

In your opinion, what is so special about Silicon Valley?

Having started my career in the Boston venture market and subsequently moved to the Bay Area, I have observed what I think makes Silicon Valley such a special eco-system: network effects. In order to grow an innovation eco-system, you need 5 scalable things:

1) Supply of entrepreneurial talent (repeat founders, enterprising college students, disenfranchised tech employees, etc.)

2) Supply of non-founder startup talent (typically, from other startups or larger tech companies)

3) Supply of true seed stage capital (angels and seed funds)

4) Supply of institutional early-stage capital (venture capital firms and corporates)

5) Large potential acquirers who value tech, talent and business growth.

There are very few geographies in the world which have the concentration of these 5 attributes at anything that approaches the network density of Silicon Valley. While startups can be built anywhere, entrepreneurial eco-systems can only flourish when they include these core constituencies.

Who have been role models you most admire?

In life, my role model is Philip Rosen (my father), the kindest, most genuine person I know who also possesses unrivaled character. In VC, I admire Matt Harris, who pioneered early-stage FinTech investing and is somehow both the nicest VC I know and one of the industry's most discerning investors.

What do you believe are the most important traits of a technology entrepreneur?

Work ethic – Building is hard.

Endurance – A startup is a marathon, not a sprint.

Intellectual creativity – Entrepreneurs face new

challenges each day. They must be able to imagine new approaches necessary to solve tomorrow's issues, not just yesterday's.

Operational flexibility – Challenges arise constantly. Inflexible entrepreneurs will quickly find themselves cornered.

Nose for talent – Building requires talent. Know who you need to build your vision and how to find them.

Strong at selling vision – Entrepreneurs must always be selling their vision (to investors, early customers, key recruits). If a founder can't sell, he won't succeed.

What do you believe are the most important traits of an early stage technology investor?

Curiosity – The best investors are always exploring and learning.

Intellectual flexibility – Venture investing requires many disciplines, dictates various working situations and demands sudden changes in direction.

Team player – It MAY be possible to invest at early stage working mostly alone, but I suspect it's much. Venture investing is fundamentally a team sport.

Conviction – If you don't have an opinion without knowing what other investors think, you are not an early-stage investor. Gather data broadly, but decide discretely.

Humble ambition – Early-stage investors will lose money. Don't let losses temper your ambitions, nor wins erode your humility.

If you could tell a technology entrepreneur just one thing, what would it be?

Expect the journey to be harder than you think and prepare for life to feel out of control often. In those times, remind yourself of the importance of your long-term mission and why you are passionate about pursuing it. After all of that, don't forget to take a deep breath and keep focused on just taking one step at a time.

Would you please create a personal quote that captures important wisdom about participating in early stage investing?

"The only certainty in early-stage investing is that plans change and the unexpected often happens. Expect surprising challenges and keep yourself open to changes in thinking. When setbacks occur, keep calm and figure out how to help your companies while remaining positive. Be sure to hold onto the learnings from these experiences, as you'll be sure to use them again in the future."

> *"Treat each investment as though it had a vanishingly small probability of achieving an extraordinarily large return."*

Earl Sacerdoti

Serial entrepreneur, Angel Investors, Advisor to early-stage tech businesses & incubators

The Copernican Group, Keiretsu Forum

What did you want to be as a child?
CEO of a tech business.

What was your first career?
Software programmer.

What other careers have you had before your current one?
Artificial Intelligence researcher, serial entrepreneur, management consultant, inventor, expert witness.

Why do you choose to focus on early stage technology investing?
I'm most excited bringing cutting-edge technologies into the mainstream, and in my experience early-stage companies do this best.

When your investing is not going well, what do you do or say to yourself to keep motivated?
Most factors affecting success are beyond the influence of investors. (Many are beyond the influence of the business and people I invest in!)

What do you love most about what you do now?
The opportunity to help others succeed.

What dreams do you have for the next 10 years?
Being able to help others while retired.

If you could change one thing about the world we live in, what would it be?
More equitable allocation of the benefits of increasing economic production to all contributors

What was your first technology investment and what happened?
After a year of planning, a group of computer scientists from a world-class Artificial Intelligence lab formed a company. Our naïve goal was to commercialize our applied research technologies in machine vision, natural language understanding, and expert systems. I invested, and joined as Director of R&D. The machine vision business ultimately attracted about $40M from angels and VCs – our lead angel investor was the founding CEO of an early Silicon Valley success and our lead VC investor was a well-respected firm that was very active in the Valley. The company developed the first industrial machine vision system, the first vision-guided industrial robot (in collaboration with a robot vendor), and the first robot programmable with a programming language (with another robot vendor). The company operated for 10 years (never making a quarterly profit) and was eventually sold to its largest customer. We funded the natural language business with a $25K Small Business Innovation Research grant that my natural-language partner and I wrote over a weekend. We spun that business out 3 years later, with additional investment from our lead VC, as a software developer and vendor. The product based on our SBIR grant generated over $400M in revenue during the following 16 years. The software business that this product initiated grew to a market value over $20 billion. The expert systems work was never commercialized, as our lead developer passed away.

Can you share a story of one of your worst investments and why it went so badly?
A first-time entrepreneur who had worked for me previously established a tech business. While I had concerns about the need to focus the business more tightly, I knew the founder well and knew he'd make the company a success. I invested and served on the Board for its first few years. The business became a modest success and has operated for over 30 years. However, I discovered that the founder had no interest in ever selling the business. My first clue might have been the identity of the CFO – the founder's spouse.

In your opinion, what is so special about Silicon Valley?

Over my 50-year career I have watched Silicon Valley evolve a highly effective ecosystem of businesses fine-tuned to support technology startups. A small group of visionaries can start a business with limited capital and even more limited personnel and bring in specialists as needed to provide expertise, services, and manufacturing and development capabilities beyond the skill set of the core team. They're often willing to work at a discount, or for no cash, in exchange for equity. The cycling of these specialists though the startup population both accelerates the refinement of good ideas and spreads them throughout the Valley.

Who have been role models you most admire?

Dr. Charles Rosen, who founded SRI's Artificial Intelligence Center and became Machine Intelligence Corporation's first CEO after he retired. Among other accomplishments, Charlie invented the power transistor, developed a way to transmit digital data over power lines in the mid-1950s, convinced the Department of Defense to begin a program to develop mobile robotics in the 1960s, and co-founded Ridge Vineyards.

What do you believe are the most important traits of a technology entrepreneur?

The ability to learn widely, comfort working with people smarter than themselves, and comfort taking action based on insufficient information.

If you could tell a technology entrepreneur just one thing, what would it be?

Don't focus on what you know how to do. You must learn many "unknown unknowns" (at least enough to enroll others to address them) to succeed. Rather than spend time in your technology comfort zone, focus on your "known unknown" problems of making the technology practical, scalable, and so on, and on recruiting advisors or staff to identify and address the "unknown unknowns". Even more important, learn how to identify your potential customers, communicate with them, and define a clear value proposition (in customers' terms) for your offering.

What do you know now that you wish you had known earlier on?

It's not enough to solve an important problem. It's critically important to identify and establish communication with job roles in target prospects whose occupants:

1. Have that important problem
2. Can admit to their management that they have that important problem
3. Have or can obtain a budget to solve that important problem.

Would you please create a personal quote that captures important wisdom about participating in early stage investing?

"Treat each investment as though it had a vanishingly small probability of achieving an extraordinarily large return."

Expect the journey to be harder than you think and prepare for life to feel out of control often. In those times, remind yourself of the importance of your long-term mission and why you are passionate about pursuing it. After all of that, don't forget to take a deep breath and keep focused on just taking one step at a time.

> *"Investing in early stage companies is like playing baseball. If you get two hits in ten times at bat), you will never get out of the minors. Three hits in ten times at bat and you will retire rich and get into Cooperstown."*

Bill Sarris

Founder, CEO and Angel Investor

Linqto, Inc., Keiretsu Connect

What did you want to be as a child?
An Entrepreneur.

What was your first career?
Real Estate Broker/Developer.

What other careers have you had before your current one?
CEO of four companies.

Why do you choose to focus on early stage technology investing?
The creativity, innovation…the juice of building something from scratch.

When your investing is not going well, what do you do or say to yourself to keep motivated?
There has to be a pony in there somewhere.

What do you love most about what you do now?
Will never have to retire. No one asked Picasso at 92 why he was still painting.

What dreams do you have for the next 10 years?
Unicorn status for Linqto.

If you could change one thing about the world we live in, what would it be?
Remove the narcissism and get back to the love.

What was your first technology investment and what happened?
Helped to create the first national online real estate listings service. Started as a touch tome dialing system. Two years later the Internet hit and it was off to the dotcom races. It was like what must have happened in the wild, wild west during the gold rush days of the 49ers.

Kids flipping hamburgers at McDonalds all of a sudden became software engineers and quadrupled their incomes overnight. The time from boom to bust was one of the most exciting in startup history.

Can you share a story of one of your best investments, and why it went so well?
It was the first fully wired, shared space in Manhattan in the Finance district at 55 Broad Street. The target market was those early incubator type companies that knew what the Internet was …circa 1995. For marketing there were only two ads placed, one in the Wall Street Journal and one in the New York Times. The copy simply said … www.55broadstreet.com … brought the exact market and filled in a few months.

In your opinion, what is so special about Silicon Valley?
If you want to make a career in finance, go to NYC. If it is fashion, try Paris. For technology there is no other place like Silicon Valley. Just look at the marketing for other high-tech locations. They will always say "it is the Silicon Valley of (another state or country)".

Who have been role models you most admire?
Scott Cook, Steve Jobs, Mother Theresa, Mahatma Ghandi.

What is the best advice you ever received about how to succeed in Silicon Valley?
Don't fall in love with your product.

What do you believe are the most important traits of a technology entrepreneur?
There is only one key to success … survival.

What do you believe are the most important traits of an early stage technology investor?
Keep you ears open and your mouth shut.

If you could tell a technology entrepreneur just one thing, what would it be?
Don't ever listen to the discouragement fraternity. No one is as interested in your success as you are.

What do you know now that you wish you had known earlier on?
Concentrate on the "big deal". The details and small stuff will take care of themselves.

Would you please create a personal quote that captures important wisdom about participating in early stage investing?

"Investing in early stage companies is like playing baseball. If you get two hits in ten times at bat (a 200 batting average), you will never get out of the minors. Three hits in ten times at bat (a 300 batting average) and you will retire rich and get into Cooperstown."

"Keep doing the best you can. Eventually things will work out."

Lilian Shackelford Murray

Secondary Venture Capital Partner

Ponte Partners

What did you want to be as a child?
Fat Cat on Wall Street.

What was your first career?
Selling bulk chemicals for Dow Chemical.

What other careers have you had before your current one?
Investment banking, money management.

Why do you choose to focus on early stage technology investing?
Actually, I/we invest in more later stage technology in healthcare. Enjoy the energy of entrepreneurs and seeing the latest innovative ideas.

When your investing is not going well, what do you do or say to yourself to keep motivated?
A mentor used to say, "You're never as great as you think you are, but you are also never as bad as you think you are."

What do you love most about what you do now?
The energy and people I deal with.

What dreams do you have for the next 10 years?
Supporting a few interesting companies to the next level and achieving balance in my own life.

If you could change one thing about the world we live in, what would it be?
Eliminate having to be on a Red or Blue team in the U.S.

What was your first technology investment and what happened?
Bought ten company stakes from a hospital chain's venture capital arm. Most of the investments were in the healthcare information space. Stepped into their shoes and had seven successful exits, two companies that were valued at nothing and ended at nothing, and one that lost a minor amount of money.

Can you share a story of one of your best investments, and why it went so well?
Purchased a stake in a drug development company that had previously had a lead drug that did not get FDA approval. Further, the company had to pull a planned IPO because of the FDA decision. We were able to purchase the stake at a very attractive price. The company subsequently resubmitted to the FDA, the product was approved, and the company had a successful IPO. Our investment was exceedingly profitable.

Can you share a story of one of your worst investments and why it went so badly?
One of my investments was in a disease management company that had multi-million-dollar contracts with the U.S. government. While the premise was good, the results were hard to measure – making it challenging to prove if the services were working or not. Ultimately, the government canceled the contracts, which led to the shutting down of the company.

In your opinion, what is so special about Silicon Valley?
It's not that the people are any smarter per se – it is the culture of being a part of something that can be big and an improvement for the future. It is ok to fail in Silicon Valley – most of the companies do. The individual just needs to keep pushing and pursuing. Ultimately, something else will work.

Why do you think the innovation economy/ Silicon Valley has a poor track record on diversity and inclusion? What needs to change?
What I have seen is a desire for excellence and all-in commitment– independent of race or gender.

There is a lot more focus now by investors, employees and customers on companies that do good (and do no harm) as well as make money. How do you think about this topic in the context of your investing?

While I think ESG (Environment, Social and Governance) has become the new buzz word for the industry, I do think that society wants to have companies/products that are well run, better for the environment and that do no harm. The market will create companies/products to meet that demand.

Who have been role models you most admire?

Venture capital firms that have successfully transitioned to younger partners while maintaining the institution.

What is the best advice you ever received about how to succeed in Silicon Valley?

Work hard.

What do you believe are the most important traits of a technology entrepreneur?

An inner drive and willingness to make sacrifices in other areas of life.

What do you believe are the most important traits of an early stage technology investor?

Ability to see trends and assess people's drive and motivations.

If you could tell a technology entrepreneur just one thing, what would it be?

Keep pushing.

What do you know now that you wish you had known earlier on?

Success comes from within.

Would you please create a personal quote that captures important wisdom about participating in early stage investing?

"Keep doing the best you can. Eventually things will work out."

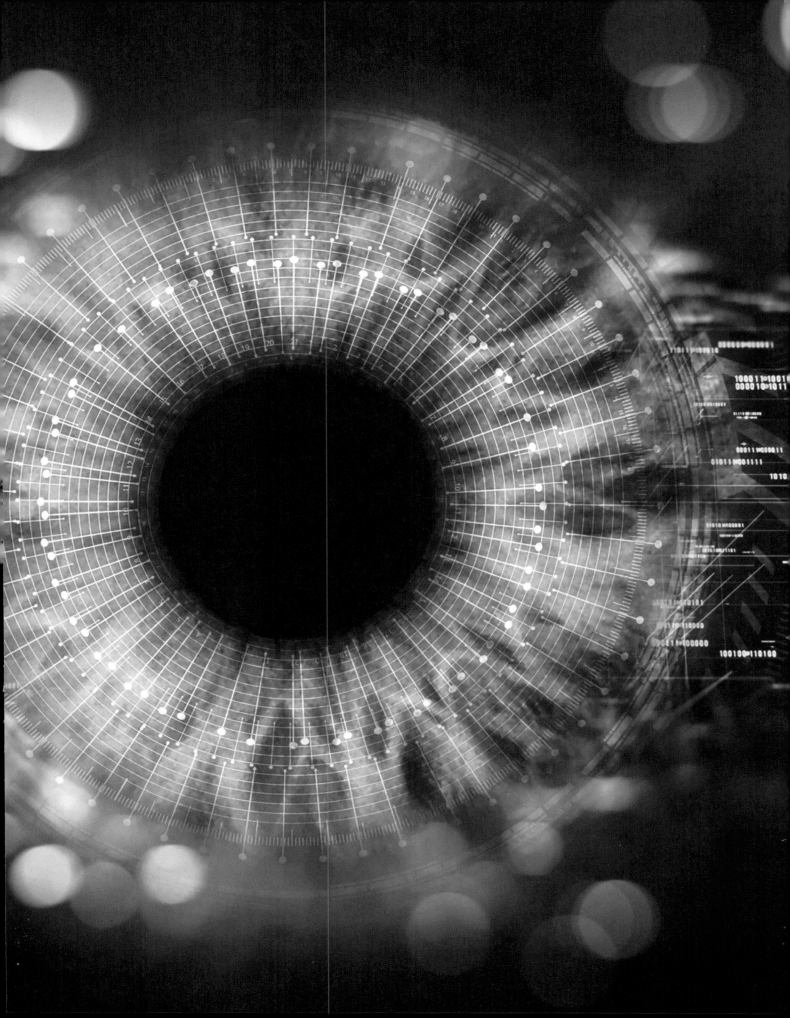

"Early stage investing is about finding, enabling, and helping founders put action behind their passion and vision. It's a team sport."

Bonny Simi

Corporate Venture Capital Investor

JetBlue Airways Senior Leadership Team, National Venture Capital Association (NVCA)

What did you want to be as a child?

When I was very young, I wanted to be a forest ranger. I lived right next to a national forest and volunteered there as a junior ranger. I even had a little outfit and everything! When I was 14, I heard inspirational speaker John Goddard, a renowned adventurer, about the 127 goals he set for himself as a teenager. He encouraged us all to go home and draw up our own lists. I wrote down five goals: Go to a good college, go to the Olympics, become a TV commentator, become a pilot, and build a log cabin. That set me on my way. I went to college initially in environmental engineering (inspired by my forest ranger experience, but eventually pivoted to aviation and became an airline pilot). Along the way, I also competed in the Olympics and worked for ABC, NBC and CBS Sports. There's one goal left for me to achieve - the log cabin.

What was your first career?

Luge Olympic athlete for Team USA.

What other careers have you had before your current one?

- President, JetBlue Technology Ventures, Captain, JetBlue Airways (current)
- Vice President, Talent, JetBlue Airways
- Director of Customer Experience & Analysis, JetBlue Airways
- Director of Customer Service, JetBlue Airways
- Director Airport and People Planning, JetBlue Airways
- JetBlue Airways Pilot
- United Airlines Pilot
- Olympic Expert Commentator, NBC-TV and CBS-TV
- Reporter, ABC-TV KGO

Why do you choose to focus on early stage technology investing?

Travel utilizes legacy technology that isn't equipped to handle evolving consumer needs. By investing in early stage technology, we're able to see the future of travel.

When your investing is not going well, what do you do or say to yourself to keep motivated?

When we're not actively investing, our focus is twofold: we concentrate both on how we can best support our portfolio companies and JetBlue.

What do you love most about what you do now?

Along with my team, we enjoy making a difference at JetBlue and the broader travel technology ecosystem.

What dreams do you have for the next 10 years?

I'm at a point in my career where I derive the most satisfaction from helping others achieve their dreams. This includes helping my team, my portfolio companies, my family, and the community to achieve new heights.

If you could change one thing about the world we live in, what would it be?

I believe that through empathy (putting ourselves in others shoes), listening, and learning, we can find win-win solutions to some of our most challenging issues.

What was your first technology investment and what happened?

Our first technology investment harnesses big data and AI to assist with airline revenue management processes. We've done several follow-on investments and have been helping them pivot their business model and test/deploy their new technology.

Can you share a story of one of your best investments, and why it went so well?

The investment that I'm most passionate about would be a lead company in the "electric air taxi" (eVTOL) industry. We invested early in this space because we strongly believe that electric propulsion will fundamentally change the aviation industry, much like jet propulsion did in the 60s. At the time of our investment, the public was still skeptical that eVTOLs would be a viable means of transportation, and there were only a handful of startups in the space. After extensive due diligence, we invested early in Joby Aviation, largely due to the strength of the leadership team and their vision. Now, not only are governments, regulators and the general public embracing the concept, but also Joby is a leading player out of hundreds of different similar startups around the world.

Can you share a story of one of your worst investments and why it went so badly?

While I won't single out any particular investment, I will say that if a venture capitalist doesn't have at least one investment that fails, they're not taking enough risks. We've had several investments that didn't pan out, but we're there to support our founders until the very last day: whether that be to try and secure more funding or pivot their business strategy. Unfortunately, you have to fail sometimes in order to learn. Fortunately for us, our successes have far outweighed the failures.

In your opinion, what is so special about Silicon Valley?

There are many wonderful entrepreneurs and successful startups from all regions of the world. What makes Silicon Valley unique is that the entire ecosystem is built around entrepreneurs to help make them successful. VCs like to invest in startups that they can work with personally, and there is a large critical mass of VCs located in Silicon Valley.

Who have been role models you most admire?

Ruth Bader Ginsburg. Early in her career, she developed a strong sense of purpose and service. She backed up her opinions with facts and data, and never took "no" for an answer. Now she stands strong at the highest court in the land and is an enduring force for good. She always puts service to others as her highest calling.

What is the best advice you ever received about how to succeed in Silicon Valley?

Silicon Valley is not as large as people think it is, and it's really built around a strong sense of community. The best advice I received is to prioritize personal relationships and not to view them as transactional. Always pay it forward.

What do you believe are the most important traits of a technology entrepreneur?

I think it's important for entrepreneurs to be fearless, but with a sense of control. Balancing confidence and humility is also key. I want to see founders with a deep sense of conviction in their idea, but also be able to ask for help and think strategically to bring their vision to life.

What do you believe are the most important traits of an early stage technology investor?

Investors should genuinely care about their portfolio companies and their founders. They need to be willing to put in the time to support their success and lend an ear when challenges arise.

If you could tell a technology entrepreneur just one thing, what would it be?

Build a team of people who are different and complementary to you. Understand your weaknesses and find people to join your team who have complementary strengths.

What do you know now that you wish you had known earlier on?

My younger self could not have imagined that I'd be where I am today. So, I would tell myself to stick with it – you can do it – and to always say thank you. You can't say thank you enough.

Would you please create a personal quote that captures important wisdom about participating in early stage investing?

"Early stage investing is about finding, enabling, and helping founders put action behind their passion and vision. It's a team sport."

In your opinion, what is so special about Silicon Valley?

If you want to make a career in finance, go to NYC. If it is fashion, try Paris. For technology there is no other place like Silicon Valley. Just look at the marketing for other high-tech locations. They will always say "it is the Silicon Valley of (another state or country)".

"Don't do it for financial return…do this because it has the promise of big returns while also being socially useful and, most of all, fun!"

Ian Sobieski

Angel Investor

Band of Angels

What did you want to be as a child?
Happy.

What was your first career?
Anti-Missile System Designer; a sort of anti-rocket scientist. Worked for the Army as part of the theater high altitude air defense program. Had some Scud missile parts on my desk as a paperweight.

What other careers have you had before your current one?
Entrepreneur.

Why do you choose to focus on early stage technology investing?
It is fun! The actual day-to-day experience is like an endless seminar series of passionate people patiently explaining technology and their plans to change the world.

When your investing is not going well, what do you do or say to yourself to keep motivated?
A couple of things: Glad I bought a bunch of real estate. Even a blind squirrel sometimes finds a nut. And you can't hit a homerun if you don't swing at the ball.

What do you love most about what you do now?
It is socially useful, personally interesting, and intrinsically productive.

What dreams do you have for the next 10 years?
To be more helpful.

If you could change one thing about the world we live in, what would it be?
Meg Zi said: "All the children who are held and loved will know how to love others; we need these virtues in the world and nothing more." Along with a felt sense of love and worthiness, if I could waive a magic wand I'd give every child the capacity for critical thinking. All combined, the world would be transformed.

What was your first technology investment and what happened?
Before I had any money, I invested my time, working for peanuts as a part-time electronics engineer for a medical device startup while still in grad school. Between classes and work I occasionally took naps under the lab bench where I was soldering electronic components. Over the course of a few months the company grew from 4 people to 40 people and switched offices twice. The experience was heady and the product promised to improve people's health and their lives. The company got a 410k and we got celebratory sweatshirts; it filed an S-1 and we started multiplying the shares we had by the IPO price; the ticker symbol popped up on e*trade and then disappeared. Our "book never filled" and the IPO was withdrawn. A downward spiral over months ended up in bankruptcy/liquidation. I now viscerally know why its important to have a "medicare reimbursment code".

Can you share a story of one of your best investments, and why it went so well?
He won the undergraduate entrepreneur contest but still couldn't raise money for his startup because the market size was too small. But this fellow was literally living in his startups lab and the size of his capital requirement was within the reach of angels. I led the syndicate that funded him and he took our few hundred thousands dollars and finished the product and built some sales. Two years later he received an acquisition offer for $7m. We had invested $300k and made 3x our money in two years; the entrepreneur had become a millionaire at 24, and his invention became part of a larger company that scaled it to nationwide use and millions of units. I've been part of better IRRs, and bigger outcomes; but this one had the good feeling that it wouldn't have happened without us as part of the team. This was 'best' because it had the most meaning.

Can you share a story of one of your worst investments and why it went so badly?
My worst investment had one of my best IRRs; and though I didn't like the lesson it taught, its an important one to remember. I lost faith in Company

X after a year. Even though I was on the board I began to feel the product simply didn't' work. The company was running out of money and the CEO continued what I regarded as frivolous spending (ie. Big money for a booth at a trade show). Giving up I resigned from the board and wrote the investment off in my book. Lo and behold, 6 months later I received an excited call telling me the company had been acquired by a major tech firm. That firm had spent more than a billion on a major acquisition and decided to buy several other companies around it as a defensive move to lock up IP and brand. That pocket change still made me 10x on this investment even though the acquirer shut down the company and never used its technology or brand. It turns out the CEO was correct about that tradeshow; a good lesson in humility.

They say its better to be lucky than smart in life, but it always bothers me that good companies can actually turn out to be bad investment and vice versa.

In your opinion, what is so special about Silicon Valley?

Silicon Valley is full of endless surprises. One can never take one's own opinion with too much certainty here. The crazy idea may turn out to change the world; the world changing idea may fizzle into nothing. The sloppy gal in a hoody is a millionaire coder; the well-dressed fellow is a fraud. It's a place of contrasts and rule breaking and norm breaking and so it serves to cause one to always question norms and rules and in so doing makes one appreciate the norms and rules to be preserved, and those to be let go of. Ultimately, its an incredibly dynamic place.

Why do you think the innovation economy/ Silicon Valley has a poor track record on diversity and inclusion? What needs to change?

We need to change our hearts and our biases. Sexism is subtle, as is racial bias. Right now I don't have many ideas and am committed to active listening.

There is a lot more focus now by investors, employees and customers on companies that do good (and do no harm) as well as make money. How do you think about this topic in the context of your investing?

One of the best things about getting older is choosing what to give a damn about. I now have the great privilege in life of paying attention to what I want to pay attention to. My new principle of investing is to invest in entrepreneurs who are doing something that is really a genuine expression of love. Nothing wrong with clever or opportunistic, but genuine passion is something want to back.

Who have been role models you most admire?

My Ph.D. professor was an amazing role of how to balance a range of interests in life. My business partner Hans taught me what it meant to be a gentleman. And my father taught me how to think and what to value.

What is the best advice you ever received about how to succeed in Silicon Valley?

Success is not about the destination, its how you comport yourself on the Journey. Do that right, and where you end up will be, by definition, a success.

What do you believe are the most important traits of a technology entrepreneur?

Hands down the most important trait is determination; the old saw that success is 10% inspiration and 90% perspiration is right.

What do you believe are the most important traits of an early stage technology investor?

Self-confidence. There is a lot of noise in the investing world and a strong tendency to follow the herd. The biggest winners (aside from those who were merely lucky) are those who came to their own decision to back something others would not.

If you could tell a technology entrepreneur just one thing, what would it be?

You are going to get more than one swing at bat. Make sure that each swing you take is sincere, but plan and behave in a way that comprehends that you're going to interact with the same people again. Treat everyone well.

What do you know now that you wish you had known earlier on?

That even in Silicon Valley the best deals rarely fall into your lap; you have to develop a thesis of what you are looking for and go find them.

Would you please create a personal quote that captures important wisdom about participating in early stage investing?

"Don't do it for financial return; it can be incredibly profitable but the odds are that your financial return will be prosaic; do this because it has the promise of big returns while also being socially useful and, most of all, fun!"

Equality of opportunity globally, irrespective of gender, race, color, citizenship, social strata, or where one is born.

> *"Leaping is best done with those that leap higher and further than you yourself can imagine."*

Jon Staenberg

Angel Investor and Venture Capital Partner

Rocketship.vc and Staenberg Venture Partners

What did you want to be as a child?
A professional tennis player.

What was your first career?
First jobs were busboy in a pizza restaurant and I also was door-to-door salesman. Of course there was also the job of "Chief Weed Picker" in the summer of my 12th year at my family's apartment buildings.

What other careers have you had before your current one?
Too many to list. From real estate to wine to car wash mogul! Restaurants to marketing. I look at the world and see so many opportunities!

Why do you choose to focus on early stage technology investing?
Now it gets interesting. The world is about to slingshot into a new realm and I love sitting in the front row.

When your investing is not going well, what do you do or say to yourself to keep motivated?
I believe. I am a true believer. Never before has humanity ever seen anything like this. It is not if, but when. So show up, really show up.

What do you love most about what you do now?
Warren Buffett says he skips to work. I feel the same way. Everyday people share their passions, dreams and show me the future.

What dreams do you have for the next 10 years?
We are at a crossroads. My dream is that we use tech and innovation to create a better world. Decisions being made in the next decade will fundamentally steer which side of the ledger humanity ends up on. We cannot sleepwalk through this and the dream is that we believe in more than the pursuit of the economic individual gains and think of how we can truly move humanity to a better place.

If you could change one thing about the world we live in, what would it be?
Compassion and Kindness have to be more integrated into our lives. The world is about to experience a rate of change that will have our heads spinning and if we don't have core human-to-human values, we will not survive.

What was your first technology investment and what happened?
Evite. I still drive the VW Bug I bought and wrapped to show my commitment! We became a proprietary eponym and were THE standard for electronic invites. I was one of the first angel investors in the deal. We had serious acquisition interest to sell for $100's of millions, a figure that would have materially changed my bank account. The larger funds had a different agenda and wanted to wait for a larger offer. It never came. There were lots of lessons. Amazing that the brand still exists decades later and yet has so little value.

Can you share a story of one of your best investments, and why it went so well?
Seagate. I was in the middle of the first dotcom bust and my fund needed a big win to overcome the carnage of the bust. I had to pivot our strategy. A good friend said I want to share with you my best idea ever. He was someone who I had always felt a strong connection with and a deep respect. I was on a cap table with some of the most noted investors in the world. That investment changed our fund and my ability to continue with later funds. You never know. Show up. SHOW UP.

Can you share a story of one of your worst investments and why it went so badly?
Amazon. This is a story of NOT investing. The check was written but minutes away from being delivered, was torn up. This would have been one of the first 10 equity checks written to the company. That check today would be worth $100s of millions. I feel lucky to even have had the opportunity but I blew it. What is amazing about this industry is the number of

opportunities to blow it but still, with one or two swings, make up for all of your mistakes.

In your opinion, what is so special about Silicon Valley?

It is like the difference of watching the event from the stands, or actually walking on the field and watching the action close-up. Silicon Valley is the field. It is where the warriors come to battle each day. Where dreams are made and broken. Where fortunes turn into movies. Where the world is changed. And like a great dynasty, Silicon Valley has the history to continue to propel it forward. There is only one Silicon Valley as there is only one Rome. Thankfully, our coliseums are still full and sitting at the epicenter of innovation and entrepreneurship for this planet. The world has tried to copy it but because of its multi-dimensionality, it is not a tracing paper exercise. But who knows what the next 10 years will bring? But I can guarantee that much of the most exciting innovation will still be coming from Silicon Valley!

Why do you think the innovation economy/ Silicon Valley has a poor track record on diversity and inclusion? What needs to change?

The Valley was started by white, male, engineers. And it truly was a club. So some of this is a remnant of that. But of course it is more complex than one simple answer. What needs to change is awareness and proactiveness. And it simply cannot be a one-time event. As the world becomes more entrepreneurial, inclusion will have to be the answer if we are to stay ahead. The old way will not win in the new world.

There is a lot more focus now by investors, employees and customers on companies that do good (and do no harm) as well as make money. How do you think about this topic in the context of your investing?

All things being equal, it adds to the attractiveness of an investment. But generally I am investing early enough, that a company has not set up its sight on doing good, it is first trying to do well! Hopefully as a coach and mentor I can help bring that ethos into the conversations and I would like to believe that I do.

Who have been role models you most admire?

Bill Gates. He taught me tenacity combined with insatiable curiosity combined with a need to win combined with confidence are ingredients for changing the world.

What is the best advice you ever received about how to succeed in Silicon Valley?

Get a real job first. Understand making decisions and having to live with them. Hiring. Firing. Selling. Cycles. And never stop learning.

What do you believe are the most important traits of a technology entrepreneur?

They should not require much sleep. And understand that you are always selling. Selling to raise money, selling to hire great employees, selling vision, etc.

What do you believe are the most important traits of an early stage technology investor?

Thick skinned, passion for learning, compassion for entrepreneurs. And what special trick(s) can you truly bring to the deal that makes you more than just more money? It is a small eco-system and I certainly do my best to try and be helpful, even if a deal is not for me. Nothing gives me more satisfaction than seeing people succeed and even more so if they do it after having failed before. But I am often told I am more transparent and honest in my feedback than other investors people meet with. And what most investors won't tell you, is that often we are surprised at what ends up being the winners. They all seem great in the beginning. Stay close, add value and be a good person.

If you could tell a technology entrepreneur just one thing, what would it be?

You are ultimately a salesperson. You sell to fund your company, you sell your products. You sell to recruit great people. And in this world, which is "noisier" than ever, be the beacon. Leadership has never been more valued.

What do you know now that you wish you had known earlier on?

How to create a path of luck by thinking bigger. I never understood how big this was all going to be. Smart money is not always smart. Every entrepreneur believes their shit. The best is yet to come. It is a marathon and one that still feels great late in the race!

Would you please create a personal quote that captures important wisdom about participating in early stage investing?

"Leaping is best done with those that leap higher and further than you yourself can imagine. And if you are going to invest in early stage companies, you are leaping. And is there anything better than catching a tailwind and an up-gust with people you love and respect? I too, am lucky to say, I skip to work everyday!"

It is fun! The actual day-to-day experience is like an endless seminar series of passionate people patiently explaining technology and their plans to change the world.

"You don't get if you don't ask."

Bart Stephens

Venture Capital Partner

Blockchain Capital

What did you want to be as a child?
Senator.

What was your first career?
Entrepreneur.

What other careers have you had before your current one?
Entrepreneur > Hedge Fund Manager > Venture Capitalist.

Why do you choose to focus on early stage technology investing?
Innovation is the heartbeat of the American capitalist enterprise. We focus on disruptive technology that will change the world in profound ways.

When your investing is not going well, what do you do or say to yourself to keep motivated?
We do more research, spend time thinking, reading and writing.

What do you love most about what you do now?
Working with young entrepreneurs, helping them to build new businesses that are creating a more fair and equitable financial system.

What dreams do you have for the next 10 years?
Blockchain technology becoming more deeply integrated into the global financial system, enabling new products and services that are available to everyone, not just Wall Street and Silicon Valley insiders.

If you could change one thing about the world we live in, what would it be?
I am currently worried about generational conflict in the next decade. Millennials currently have 1/7th the wealth boomers had at that age. There is less home ownership and they earn 20% less than boomers at the same age. In developed nations spending policies around healthcare, entitlement programs and future pension fund insolvencies will cause more bailouts benefitting boomers at the expense of millennials and Gen Z. The difficult decisions ahead of us will likely be decided legislators that are boomers themselves. Nobody wants to steal from younger generations but that is effectively what is happening on our current trajectory.

Can you share a story of one of your best investments, and why it went so well?
Buying Bitcoin at less than $100 a coin. We learned about Bitcoin from an entrepreneur who was fascinated by the Bitcoin protocol that allowed for value exchange over a blockchain. In order to best understand the technology, we started assembling our own Bitcoin mining computers in our Father's garage in the hills of Tiburon. We soon blew out all of the transformers at the house and his power bill went up 10x, but understanding the hardware aspects of Bitcoin mining gave us a better appreciation that Bitcoin was just the tip of the iceberg and in the next 20 years we would see thousands of blockchains addressing all types of industries that the story was bigger than just Bitcoin. Eventually, we went on to found the Venture Firm Blockchain Capital to focus on investing in the companies that were enabling effective and parallel financial systems. This system allows for a more inclusive form of capitalism and turns many of the institutions that we are used to on their head. In a blockchain ecosystem, the users are the owners and those owners become advocates. This is a new powerful model of capital formation and arguably a new form of capitalism that I call crypto communal capitalism.

In your opinion, what is so special about Silicon Valley?
Silicon Valley is often thought of as a location, but I prefer to think of it as a mindset. Entrepreneurship is about risk taking and dreaming big. Many of the local universities foster great computer science departments and engineering departments that enable young founders to feel comfortable taking significant risks early in their career. The Founders are geographically close to Venture Capitalists and other investors that share that mindset. It is a mindset that focused not only on risk taking and dreaming

big, but thinking creatively about how technological innovations can improve the human condition in dramatic ways. Silicon Valley positioned as the nexus of these values is not assured. In the blockchain technology industry, we are seeing innovation happen around the globe and especially in Asia. One of the things that is important to maintain Silicon Valley leadership is support by government institutions, including tax policy, regulation, and immigration. For example, an uncertain regulatory environment in the United States surrounding crypto assets, is effectively forcing innovation offshore to China, Singapore, Switzerland, and other jurisdictions that are eager to import the Silicon Valley mindset and work ethic.

Who have been role models you most admire?

My father, Paul Stephens, has been a foundational business role model for me. His firm, Robertson Stephens, financed thousands of Silicon Valley companies throughout the 1970s, 1980s, 1990s and we started talking about disruptive technology around the dinner table starting at age 10.

What is the best advice you ever received about how to succeed in Silicon Valley?

Brook Byers, one of the founders of Kleiner Perkins Caufield & Byers, shared with me some insights that have always stuck with me. To be successful as an investor in Silicon Valley, it is not enough to identify early important technologies and that sometimes the best investments are made in a contrarian fashion, when you are going against the grain of conventional wisdom around a given technological development. The importance of timing is also critical. Netflix was a great business idea that was tried multiple times before widespread broadband penetration enabling its eventual inflection point and success.

What do you believe are the most important traits of a technology entrepreneur?

The best entrepreneurs are not simply trying to build a new product or service, but rather building a mission driven organization. Employees at a startup want to be a part of something that is bigger than pursuing their own self-interest, so it is important to build a culture built around common ideals that pursue a larger mission.

What do you believe are the most important traits of an early stage technology investor?

It is incredibly important to be a good listener. Many Venture Capitalists that have achieved success can easily fall into the trap of pattern recognition, similar to generals fighting the last war. By listening to entrepreneurs and providing candid and transparent feedback early in the growth of a company, you can establish a relationship that can endure the inevitable challenges and setbacks that all startups face.

If you could tell a technology entrepreneur just one thing, what would it be?

Don't die. It always surprises me how much founders worry about competitors when their primary mission is to ensure the survival of all of their stakeholders: their employees, customers, and shareholders. The #1 job of a founder is to ensure that he has enough capital to pursue her mission.

What do you know now that you wish you had known earlier on?

After being an entrepreneur and Hedge Fund manager and a Venture Capitalist for 20 years, I have a better appreciation for the role of timing and luck. The best teams, products and entrepreneurs sometimes do not arrive at the market at the right time, regardless of the talent or work ethic of the founders or the engagement of the investors. Much of the success of the company is dependent on factors that are outside of the control of the founders and the investors. This is an uncomfortable truth, but the best team does not always win the game.

Would you please create a personal quote that captures important wisdom about participating in early stage investing?

"You don't get if you don't ask. Entrepreneurship is really hard work and it is a team sport. Founders should ask for as much help and mentorship as they can from their investors and other stakeholders. People are often willing to help founders out but they usually need to be asked."

It's not that you are encouraged to think big, it's that you are expected to do so. The American Dream in the making is always on display and it conveys a sense of optimism and hopefulness, even while also demonstrating its own silly self-centeredness.

"It's so essential not to get jaded because you've "seen everything;" you must always preserve your sense of wonder and your ability to laugh at yourself."

Lisa Suennen

Venture Capital Partner; previously an Entrepreneur

Manatt, Phelps & Phillips, LP and also Venture Valkyrie, LLC

What did you want to be as a child?
Investigative journalist, having seen All the President's Men.

What was your first career?
Tech marketing & public relations.

What other careers have you had before your current one?
Entrepreneur/healthcare services operations/sales/marketing.

Why do you choose to focus on early stage technology investing?
There is nothing more exciting than to be a part of new ideas as they evolve from their drawing board to household name.

When your investing is not going well, what do you do or say to yourself to keep motivated?
There are always cycles – bad things happen, but good things happen too.

What do you love most about what you do now?
I get exposed to a diversity and creativity of ideas from incredibly smart people – it's a rush.

What dreams do you have for the next 10 years?
To stay in the thick of new ideas; to invest in more things that solve big problems, even if doing so may be less profitable than other targets; to live by the water and write a book.

If you could change one thing about the world we live in, what would it be?
I would figure out ways to incentivize people to express more empathy, kindness and consideration about others and that includes those of different nationality, race, gender equality and political persuasion.

What was your first technology investment and what happened?
My very first investment was in a remote patient monitoring software-company for following people with chronic diseases to prevent major adverse medical events. It was designed for ease of use and simplicity for the patient and simplicity of workflow by care managers. The company exited basically for IP value and the investors made little. The challenge was that the company was too ahead of its time – it would soar today—and didn't understand that you could not simply apply tech approaches to the healthcare market without a real understanding of reimbursement and insurance incentives. Lessons learned: follow the money and tech rules don't apply in healthcare.

Can you share a story of one of your best investments, and why it went so well?
While it hasn't exited yet, I think Evidation will prove to be one of my most valuable and gratifying investments. The valuation has grown dramatically, but entirely commensurate with the real value this real-world evidence/patient data collection company has demonstrated. Founded and led by women, they have been smart, quick to adjust, kind and effective leaders. This is a young company that works with the biggest of the big companies in healthcare and keeps itself lean and always innovating. I am excited to see where it goes.

In your opinion, what is so special about Silicon Valley?
It's not that you are encouraged to think big, it's that you are expected to do so. The American Dream in the making is always on display and it conveys a sense of optimism and hopefulness, even while also demonstrating its own silly self-centeredness.

Why do you think the innovation economy/ Silicon Valley has a poor track record on diversity and inclusion? What needs to change?
People tend to collect others like themselves and really don't understand and appreciate the incentives inherent in diverse leadership, team building, and

product design. It's a shame because they are leaving money on the table for themselves, which should be a powerful motivator to do the right thing. People are remarkably incapable of seeing what is right in front of them.

There is a lot more focus now by investors, employees and customers on companies that do good (and do no harm) as well as make money. How do you think about this topic in the context of your investing?

I think about it all the time. It is a core tenet of healthcare investing, where I spend most of my time anyway, but the field could be even more intentional about it. In my opinion, we can do better here too, as we must stop ignoring innovation in children's medicine because "the market is too small." Healthy children lead to healthy adults, and our nation depends on our improving this equation. It's good to see other fields starting to think about this more. Shouldn't products that improve our lives and do no harm be worth more?

Who have been role models you most admire?

My father, entrepreneur and investor Albert Waxman

Women VCs who came before me

Sue Siegel, my former boss at GE Ventures

Shannon Kennedy, my former boss at American Biodyne

What is the best advice you ever received about how to succeed in Silicon Valley?

Be yourself; it's ok if not everyone likes you.

What do you believe are the most important traits of a technology entrepreneur?

Agile, thoughtful, fair, ethical, resilient, terrified of running out of cash; even more terrified of underspending to the point it impedes the goal; understanding that people need to have lives outside of work; able to admit mistakes and ask for help; clear-eyed, transparent and self-reflective; good sense of humor

What do you believe are the most important traits of an early stage technology investor?

Patient; responsive; possessed of high expectations and big aspirations; demanding but kind; willing to give credit where due; humble; committed to diversity in action, not just words; understanding about people's need to have lives; always thinking about ways to make the company more successful.

If you could tell a technology entrepreneur just one thing, what would it be?

If you don't ask, you don't get.

What do you know now that you wish you had known earlier on?

You need to be your own best friend and advocate.

Would you please create a personal quote that captures important wisdom about participating in early stage investing?

"It's so essential not to get jaded because you've "seen everything;" you must always preserve your sense of wonder and your ability to laugh at yourself."

What do you love most about what you do now?

Fostering innovation in startup hubs around the world and promoting free market entrepreneurship, the most positive and transformative economic activity in world history.

"There are lots of good ideas, but making a successful business out of any of them is really hard."

Clifford Tong

Angel Investor

Berkeley Angel Network, Sand Hill Angels, Insync Angels, Gaingels

What did you want to be as a child?
Athlete (baseball, basketball, tennis, golfer), then an entrepreneur after I figured out I wasn't good enough, which didn't take long.

What was your first career?
Marketing/sales in data processing (that's what we called it back then).

What other careers have you had before your current one?
Management consulting, because I was better at telling others what they were doing wrong than knowing what I was doing wrong.

Why do you choose to focus on early stage technology investing?
Because at the end of the day you want to be able to say you tried to make a difference.

When your investing is not going well, what do you do or say to yourself to keep motivated?
I ask myself whether I at least had fun and did I make the right decision with the information I had at the time.

What do you love most about what you do now?
I get to learn about things I would never get exposed to and get involved with activities I would never get to do.

What dreams do you have for the next 10 years?
That our society will address our most serious problems, namely wealth disparity/income inequality, systemic racism/social injustice, money in politics, gun violence, police corruption, access to healthcare, environmental sustainability.

If you could change one thing about the world we live in, what would it be?
That people would be more self-aware and realize that the world does not revolve around them.

What was your first technology investment and what happened?
I invested in a startup electric vehicle company in the 90s that was way ahead of its time. They eventually ran out of money and could not raise the capital necessary to bring the car to market, despite a significant amount of visibility and high-profile investors. What I learned is that well-established industries like the auto industry are often opposed to change and entrenched companies have a strong "not invented here" bias, although Tesla has since been a welcome change.

Can you share a story of one of your best investments, and why it went so well?
I invested a startup telemedicine company in 2006 because I thought it would be the future of healthcare. While they did have some early traction, it was still an idea ahead of its time and they floundered for a while. The turning point was bringing in a new CEO who had significant M&A experience and connections and he was able to get new investors to inject the growth capital needed. They eventually went IPO and are the industry leader.

Can you share a story of one of your worst investments and why it went so badly?
I invested in a removable disk storage startup back in the 90s that had a really innovative product a month after I invested, I could have sold any or all of my stock to a new investor for an 8x return. Unfortunately I decided not to sell any of my stake in the company and eventually lost it all. I have since learned not to be so greedy and when presented with the same opportunity, sold enough to recoup my initial investment so the rest was gravy.

In your opinion, what is so special about Silicon Valley?
New ideas are expected, not frowned upon. The status quo is always questioned, with an eye towards improvement, even if it means breaking the mold. There are no sacred cows. No one ever says something has to be that way because we have always done it that way. We welcome that as an opportunity, not a roadblock.

Who have been role models you most admire?

Barack Obama, because he guided our country through turbulent times and always knew the right balance to strike and which battles to fight and when. He is not an entrepreneur but we can learn from him about what it takes to be a leader, and focus on the big problems instead of sweating the small stuff

What is the best advice you ever received about how to succeed in Silicon Valley?

"Genius is 1 percent inspiration and 99 percent perspiration" - Thomas Edison.

True then, true now.

What do you believe are the most important traits of a technology entrepreneur?

I think you want someone who has vision and imagination, but not afraid to change it if the realities dictate it. And they must be execution-oriented and able to get things done in order to see that vision come to fruition.

What do you believe are the most important traits of an early stage technology investor?

Do your homework, and don't be afraid to walk away if a deal isn't as interesting after taking a closer look. We tend to get excited about a deal early on, and then ignore red flags we find in due diligence.

If you could tell a technology entrepreneur just one thing, what would it be?

Make sure you are solving a significant problem with compelling interest in your solution. If your solution is a "like to have", not a "need to have", that is not good enough and you should look elsewhere.

What do you know now that you wish you had known earlier on?

That the main reason startups fail is lack of a strong enough market for the product, so it makes sense to determine whether there is compelling interest for a market solution up front, which is not really that hard to do. If the entrepreneur hasn't proven this to us, we shouldn't invest.

Would you please create a personal quote that captures important wisdom about participating in early stage investing?

"There are lots of good ideas, but making a successful business out of any of them is really hard."

> *"Be as passionate about growing your business as you are about building your product."*

Ron Weissman

Angel Investor and former Venture Capital Partner

Band of Angels and Angel Capital Association (Board Member)

What did you want to be as a child?
US Senator.

What was your first career?
Professor of Medieval and Renaissance History.

What other careers have you had before your current one?
Five prior careers, including university professor, VP of Academic Computing, consultant for various federal agencies and private sector companies, CMO (including CMO and head of investor relations at Verity); head of Strategic Marketing and European Marketing for NeXT, reporting to Steve Jobs, venture capitalist with Apax Global (2000 – 2016).

Why do you choose to focus on early stage technology investing?
I am passionate about innovation. Early stage, while not the only way to foster innovation, is a critical one.

When your investing is not going well, what do you do or say to yourself to keep motivated?
What can I do to help? Remember to think long term and take a portfolio-wide view and learn from successes and failures. Both count!

What do you love most about what you do now?
Fostering innovation in startup hubs around the world and promoting free market entrepreneurship, the most positive and transformative economic activity in world history.

What dreams do you have for the next 10 years?
Doing my bit to reverse the ongoing decline of our civilization.

If you could change one thing about the world we live in, what would it be?
I would return education to its classical liberal roots and emphasize free inquiry, the scientific method, the marketplace of ideas, diversity of thought and freedom of expression. I would replace today's emphasis on indoctrination in a rigid set of illiberal orthodox views with the values of a truly liberal education.

What was your first technology investment and what happened?
My first technology investment, a dotcom vintage mid-stage deal, did not go well. The startup, an emerging category leader, brought top-notch tech support and Internet enablement to small and mid-sized businesses similar in quality to the services provided by system integrators serving G2000 companies. Impressed by the partnerships the company had formed with global market makers across every facet of its business and the charisma of its founders, we overlooked basic risk factors: a very high valuation, a customer base consisting mostly of fragile startups and non-profits at great risk from economic down cycles, a management team that was brilliant but not very flexible about pricing and an all-in-one, take it or leave it solution. The deal closed before the dotcom collapse. Shortly after, the dotcom bust eliminated almost 100% of the company's installed base. And larger prospects were unwilling, post-bubble, to trust critical IT resources to startups. The company survived for a decade but was eventually sold for pennies on the dollar.

But the lessons learned were invaluable: realistic valuations are essential, partnerships can be overhyped, diligence on all the business levers underlying financial viability is a must and be very, very wary of charismatic teams. The dotcom bust exacerbated the firm's core problems, but did not cause them. My firm had a saying, "beware your first deal." And they were right.

Can you share a story of one of your best investments, and why it went so well?
This investment (a team deal), was the reverse of my first investment in every respect. Rather than being in a hot, sexy field, this was an infrastructure

software provider, doing un-sexy things: integration of corporate data systems. Founded by a brilliant technologist, we eventually hired a CEO who both a leader in integration technologies and a skilled business leader with strong business skills learned in the trenches. He had founded and managed a similar company to a very successful exit. As a CEO he had far more wisdom and common sense than charisma, and was a great "customer-whisperer." He listened with at least as much intensity as he pitched his product. Ultimately, he became a trusted advisor to leading customers world-wide. The company exited very profitably after a sale to a market maker. The reasons for our success: 1) The CEO was pragmatic, skeptical, ruthlessly honest, not afraid to challenge company sacred cows, including his own sacred cows. 2) The CEO/company hired well and expected excellence across every functional areas. And the CEO had a great technical network to hire from. 3) The company prized customer intimacy and never thought it was smarter than its users. Product/market fit was an overriding goal. 4) The Board worked well together and brought the company large dollar OEM partnerships with industry leaders. 5) The company did many minor pivots to ensure it was meeting high-value customer needs. 6) The company was quite flexible in its business deals and turned services into a competitive weapon.

Can you share a story of one of your worst investments and why it went so badly?

A healthcare company providing IT support to small physician practices, reacting to the requirements of the Affordable Care Act. The company became a market leader, which, after successive rounds, had one out of every six US residents in its system and achieved a near-unicorn valuation. But the company prized growth over profitability. After many years, the market changed in two unfortunate ways: 1) A tougher fundraising climate made investors prize profitability over topline growth and the company was still unprofitable. 2) The health IT market changed. The target market, small physician practices, was no longer perceived as attractive, given pressures on small practices to merge with large care networks. Unfortunately, company's products were not designed to meet the requirements of large healthcare providers. The company was eventually sold for a fraction of its prior valuation. Last round investors (like myself) did make some money due to our position in the preference stack, but not enough

to recoup the entire investment due the very large preference stack senior to our primary position as early round investors. (We invested very early, and then very late). Key takeaways: 1) market sentiment can change dramatically over the life of a long-term startup investment. 2) Understanding future likely capital models is critical for early stage investors so as to understand later stage capital risks, particularly the risk of very large liquidation preferences stacks. 3) In times of stress, most markets which have forsaken the laws of financial physics ("get big fast, don't worry about profits" will find that the old laws reassert themselves with a vengeance.

In your opinion, what is so special about Silicon Valley?

Our rich ecosystem still surpasses startup hubs around the globe. It took 30+ years to develop Silicon Valley. You can't will an ecosystem into being and governments can't build one by decree or bags of cash. It must develop organically or it will be fragile and incomplete. It takes time to cohere, to work together and build trust.

Our ecosystem includes universities, research labs and large tech and biotech companies who have created or spun out hundreds of market leaders. We have the world's largest community of angels, VCs, CVCs, incubators and accelerators. Also, public markets, venture law firms, accounting firms and Investment banks that monetize our success, and executors and undertakers who turn around or wind-down our failures.

Repeat entrepreneurs and experienced technical and executive talent and recruiters are key, too. In much of the rest of the world, when companies need serious capital, they move to Silicon Valley, depleting local human capital. Our talent is already here, and, historically, each new technology wave brings us more talent.

But there are dangers: 1) We are narcissistic. We're not the center of the universe. 2) Ours is a political monoculture that punishes diversity of political or social thought. 3) Our cost of living is too high to sustain our growth. 4) Silicon Valley is Hollywood for smart people. Unfortunately, some CEOs love being celebrities, creating a repulsive corporate culture. And our cult of the CEO leads to egomania, to scandals like Theranos and all too prevalent sexual harassment. We're supersizing the worst of contemporary culture, given our outsized influence.

Let's bring back the humility, pragmatism, tolerance and sense of wonder of the Valley's early years.

Who have been role models you most admire?

My father (who helped build Kaiser). John Huston (Ohio Tech Angels). The Good Steve (Jobs), not the Bad Steve. Cosimo de'Medici, Renaissance Florence's most influential citizen. Peter Thiel (for repeatedly challenging orthodoxy and enabling talented kids to do something far more important than college.) And many successful, humble CEOs you've never heard of.

What is the best advice you ever received about how to succeed in Silicon Valley?

Be fact-driven. Try to give anyone who approaches you something: a network contact, an honest view of their business, a potential peer who can help build or ground their business.

What do you believe are the most important traits of a technology entrepreneur?

- Being very intelligent
- Having a unique market insight
- Knowing what you don't know and seeking to fill those gaps
- Hiring well, "hiring different," and being a good judge of people
- Being humble, absorbing criticism well
- Dealing well with ambiguity
- Ability to focus on what's most important.

What do you believe are the most important traits of an early stage technology investor?

- A good judge of people
- Intelligent, intuitive and analytical
- Ability to think outside the box, including the investor groupthink box
- Honest and humble
- Risk tolerance
- A willingness to be hands-on and help operationally—if asked

- A character strong enough for the early stage roller coaster and standing for what's right even in conflict with fellow investors
- A curious and voracious consumer of information
- A good mentor, who does so with honesty and empathy
- Being a coach, not a cheerleader
- Having enough operational experience to overcome the prejudices and disabilities which come with an MBA
- Not being an ego-driven Master of the Universe jerk.

If you could tell a technology entrepreneur just one thing, what would it be?

Please be as passionate and insightful about building your business as you are about building your product. Experienced CEOs and investors care much more about building a minimum viable business (MVB) than a minimum viable product. (MVP).

What do you know now that you wish you had known earlier on?

It is the boring stuff that matters. The boring stuff may make the difference between investing success and failure. It may be important to have a sexy story and vision to attract investors' attention, but beyond that investor returns are gated by financial models, investor and company capital models, valuation, customer insights, and detailed focus on metrics. My first deal was all about vision, charisma, a compelling story. My successful deals have all been about metrics, milestones, honesty and effective execution—the very things that bore and frustrate some visionaries.

Would you please create a personal quote that captures important wisdom about participating in early stage investing?

"Be as passionate about growing your business as you are about building your product."

"Be an independent thinker. And when you find your thoughts have led you to a contrarian position, that's where the opportunity lies."

Colin Wiel

Angel Investor, Tech Entrepreneur

Mynd

What did you want to be as a child?
Entrepreneur.

What was your first career?
Engineer at Boeing.

What other careers have you had before your current one?
Engineer, entrepreneur, real estate investor, angel investor.

Why do you choose to focus on early stage technology investing?
I love thinking about how technology is changing the world, and helping entrepreneurs achieve their dreams. And, I get great returns.

When your investing is not going well, what do you do or say to yourself to keep motivated?
I focus on my long-term track record, which is very good, and remind myself that there have been plenty of ups and downs along the way.

What do you love most about what you do now?
I love the perspective I get wearing both hats at the same time: being an entrepreneur and continuing to back other entrepreneurs.

What dreams do you have for the next 10 years?
I am living my dream, launching companies and helping others do the same. I would love to see my company Mynd really transform the residential investing world and open up that asset class to lots more people. Both inside and outside of Mynd, I want to create ever-improving software models of human decision-making. While AI does an incredible job of solving many problems, it does not do a great job of mimicking many decisions that human experts make easily.

If you could change one thing about the world we live in, what would it be?
I would reduce the population from 7.8 billion to 2 billion people. Then our planet could sustain us all with a middle-class or better standard of living, with plenty of space to also sustain the biodiversity that we still have left.

What was your first technology investment and what happened?
Microsoft, in 1990. I was just out of college and opened my first brokerage account. All I bought was Microsoft stock. This was early days for Microsoft, before Windows and Office were released. For two years, every time I saved more money, I bought more Microsoft stock. It was the only stock I owned. There was no diversification, but I was so certain that it was going to keep going up that I didn't care. It went up 100x within 10 years. Back then companies went public much earlier, so it was possible to get into them when there was still a ton of upside to be realized. Since then, they started waiting much longer to go public, which one reason I turned to angel investing.

Can you share a story of one of your best investments, and why it went so well?
One of my best investments was Brightroll, a video ad network started by my friend Tod Sacerdoti. I invested in 2006, and exited five years later when it was sold to Yahoo for $640 million. I made a 55x return. My reasons for investing were twofold. First, I had played soccer with Tod for years. You can tell a lot about a person's integrity and grit on the soccer field, and Tod was top notch on both counts. It was largely a bet on Tod. Second, when investing in technology, I always try to look at where the world is moving, and what is coming next. Like Wayne Gretsky famously said, he skates to where the puck is going to be, not to where the puck is. In the 1990's, websites consisted of text, colored backgrounds, and simple graphic images. By the early 2000's, as bandwidth continued to grow exponentially, pictures became a core part of the web experience. By 2006, videos had started to appear on the web. This was early days for video – Youtube had not yet been acquired by Google – but if you followed the exponential curve of bandwidth growth, it was clear that video was going to become more and more important on the web. I always look for that stage of technology, where it is just about to have a big impact on our lives.

Can you share a story of one of your worst investments and why it went so badly?

One of my worst investments was not in technology, but was a fund that would buy, renovate, and rent out apartments in a long-term buy-and-hold strategy. I love this asset class, and have done very well with it overall. This asset class tends to be very safe, reliable, and tax-efficient. However, in this particular case the operator turned out to be unethical, and had a scheme to evade rent control in the city in which it was focused, which he did not disclose to the investors. He ended up getting sued by the city, and lost everything. I had a bad feeling about this investment in my gut, but rather than thinking for myself and doing my own due diligence, I followed other smart investors that I knew. They were in turn following other people they trusted, who were presumably doing the same. I am not sure any of the investors actually did any due diligence. My takeaway from this experience is to always think for myself, do my own due diligence, and trust my gut.

In your opinion, what is so special about Silicon Valley?

Ever since the gold rush, the Bay Area has attracted adventurers, risk-takers, and freethinkers. In the 1960's, the free speech movement was centered in Berkeley and the Summer of Love, in San Francisco. Human nature is to follow the beaten path, and to stick with what has been proven to work. However, technology is all about change, about doing things that have never been done before. To be successful, one has to go against human nature and think outside the box. That is what the Bay Area fosters. To me, this is epitomized by the famous 1984 Apple Macintosh Superbowl commercial, an allusion to George Orwell's 1984, showing a young woman chased by guards running past grey-clad zombie-like men listening to a "big brother" figure on a giant TV screen. She throws a sledgehammer through the screen causing it to explode, with the narrator promising that "1984 will not be like 1984". The freethinking culture of the Bay Area is a self-fulfilling prophecy, drawing in free-thinkers and repelling conservative thinkers in a never-ending virtuous cycle.

Why do you think the innovation economy/Silicon Valley has a poor track record on diversity and inclusion? What needs to change?

Implicit bias is an evil force that permeates all aspects of our society, making it deeply entrenched and difficult to change. It exists in schools, with the police, with hiring managers and store clerks and landlords. There are interdependencies and feedback loops between all these different facets. The only way we can create meaningful change is to address all these facets concurrently. Change is happening, but much too slowly. You have to zoom out to a time scale of decade-to-decade to be able to see the change. We should be able to see change looking year-to-year. This problem is especially tricky because even the best intentioned of us do things subconsciously that perpetuate the problem, such as The George Floyd protests, are helping bring this closer to the top of our consciousness.

There is a lot more focus now by investors, employees and customers on companies that do good (and do no harm) as well as make money. How do you think about this topic in the context of your investing?

My primary goal in life is to have a positive impact on the world. I am somewhat skeptical about non-profits, so I largely look to have that impact through my investing.

Who have been role models you most admire?

Bill Gates. He is possibly the most intelligent person on the planet, and was able to harness that intelligence and have a massive positive impact on the world.

What do you believe are the most important traits of a technology entrepreneur?

Having extreme confidence in one's own vision, and the charisma to entice other people to believe in that vision; and, being right.

What do you believe are the most important traits of an early stage technology investor?

Being a first-principles thinker, being social and likeable, and having a good sense for money.

If you could tell a technology entrepreneur just one thing, what would it be?

Take more chances - make more mistakes.

What do you know now that you wish you had known earlier on?

Crawl, then walk, then run. Don't try to start out running.

Would you please create a personal quote that captures important wisdom about participating in early stage investing?

"Be an independent thinker. And when you find your thoughts have led you to a contrarian position, that's where the opportunity lies."

"Embrace globalization and be two steps ahead of everyone else."

Klark Xia

Founder, Angel Investor, Advisor

Bowei Blockchain Technology, Fifth Era Blockchain

What did you want to be as a child?
I wanted to be a strategist, a tactician behind a successful organization. This idea was planted in my mind when I read Romance of the Three Kingdoms at a young age as I was deeply fascinated by the wisdom and tactical thinking Zhuge Liang displayed throughout the story.

What was your first career?
I started my career in the video game industry as a Game Analyst. My job was to help the game design team make sense of the player behavior in games and make appropriate changes to the games for better user experience and monetization.

What other careers have you had before your current one?
Before working at Fifth Era Blockchain and founding Bowei Tech, I was the Director of Publishing at Concept Art House. I led the publishing division of the company where we localized, optimized and published a portfolio of successful western mobile games to the eastern markets and vice versa.

Why do you choose to focus on early stage technology investing?
Changes are exciting, and I love that early stage tech startups can quickly go from an idea to bringing changes to our daily lives. I love seeing exciting changes first-hand!

When your investing is not going well, what do you do or say to yourself to keep motivated?
Modern psychology suggests that humans remember failures better than successes – that's great, because each vivid detail I remember from my failures can potentially fuel my next success.

What do you love most about what you do now?
I get to regularly meet successful entrepreneurs and investors who each provide unique perspectives. Their experiences help me make better decisions for my own businesses.

What dreams do you have for the next 10 years?
For the next 10 years, I'd like to see blockchain technology have meaningful applications everywhere in our lives. I want to see governments, large corporations and non-profit organizations use the technology to fix the existing security loopholes like vulnerable centralized storage of confidential data or to reinvent the outdated systems such as the international payment system. I invested both money and time into building businesses around this technology to hope my dream can come true one day and change certain areas of our lives for better.

If you could change one thing about the world we live in, what would it be?
The world's pandemic response system needs to change. Even with the modern technologies available, the current pandemic response system is still very weak. In the battle again COVID-19, we see major countries each fighting its own uphill battle rather than forming a united front against a novel virus. This needs to change – countries need to share medical data with each other (blockchain can help do that without giving up privacy), governments need to better support the frontline medical staff, information about the disease should be relatively consistent across different countries. The list goes on, but the point is that when facing a pandemic that's killing people every day, we need to fight it together.

Can you share a story of one of your worst investments and why it went so badly?
My worst investment was a growing company with a good idea suddenly went under due to its CEO's bad money management and fundraising skills. The CEO kept telling the investors everything was going well and the business was growing every month until one day, all the investors received an abrupt email from the CEO one day announcing the company ran out of money and was dissolving in a couple of days. Most of the investors were not aware of the money issue until they received the email.

The truth of the matter was, the executive team failed to convert the cryptocurrency they raised into cash as soon as they could, so they were left with a third of the money raised when the crypto bear market struck. The CEO was too afraid to tell all the investors so he only told a few investors he's closer with and asked for additional funding from them. One of those investors agreed but eventually withdrew right before the company ran out of money. By then, it was too late for the CEO to ask other investors to consider providing additional funding to save the business.

This failed investment now serves as a constant reminder for me, I specifically and regularly ask about the financials of my portfolio companies and explicitly ask the executives to inform me immediately if the companies run into issues that may threaten the businesses' future.

In your opinion, what is so special about Silicon Valley?

Silicon Valley to "tech people" is like Hollywood to "film people," it gives everyone two special things: dreams and pressure. Silicon Valley brings some of the brightest minds in the world together, and cool things happen when great people work together. It gives entrepreneurs a pool of high-quality candidates to build a great team and an abundance of investors who are willing to take the chance for the "next big thing." As an entrepreneur, you only dream bigger when you realize that you can have the tools and the people to help bring your ideas to life. Inevitably, the concentrated talents in a relatively small area create the toughest competitions. No matter what role you play in the early stage ecosystem, you're constantly facing the highest level of competition in nearly every sector. As cruel as it may sound, the constant competitions and the US work Visa system combined are pruning the populations in the space, practically only allowing the top-performing individuals to stay in this land of opportunities.

Who have been role models you most admire?

I was still in middle school when our family went from middle class to hundred of thousands of dollars in debt after a number of failed start-up attempts and investments. My mother displayed a great deal of intelligence, confidence, and resilience, and showed me what it means to never give up.

What is the best advice you ever received about how to succeed in Silicon Valley?

"It's never going to be easy in Silicon Valley, just know there'll be tough competitions and face them head on." I got this advice when I was 17, right before I came to Silicon Valley for education. It was important for me to have the realization that I was going to face competitions and pressure early on, so I was never really surprised by any challenges I faced since. I try not to chase after success, instead I work to get better and make sure I'm always in a challenging environment.

What do you believe are the most important traits of a technology entrepreneur?

I believe every entrepreneur needs to be a good storyteller with the ability to live up to the told stories. A good storyteller should be able to convince the investors that you are "the one" to lead the company and come out victorious against the competitions; she should be able to convince her employees that their stock options will eventually be worth more than the higher salary bigger companies are probably offering them; and most importantly, a good storyteller should be able to remind herself why she started the business and why she should always believe in herself when she inevitably face difficult situations at some point in the future.

If you could tell a technology entrepreneur just one thing, what would it be?

Expand your international network! Technology trends usually start in Silicon Valley and quickly spread to the rest of the world, which makes it a wonderful opportunity for entrepreneurs who started their business early in the wave. If you have an extensive international network, you can potentially expand your business to other countries when they catch on the hype. At that point, the domestic market might already be saturated with the same type of startups and investors have already picked their horses, but you would still be able to gain interests from clients and investors from abroad.

Would you please create a personal quote that captures important wisdom about participating in early stage investing?

"Embrace globalization and be two steps ahead of everyone else." As an early stage investor in Silicon Valley, you usually get to see the "new tech trends" a few weeks or even months earlier than the foreign investors, which gives you the advantage to take the observations and experience to other countries and invest in the leaders there before they get "hot." Taking advantage of the language and cultural barriers between Silicon Valley and the rest of world has brought success to many investors - it's never too late to start learning another language."

Ever since the gold rush, the Bay Area has attracted adventurers, risk-takers, and freethinkers. In the 1960's, the free speech movement was centered in Berkeley and the Summer of Love, in San Francisco. The freethinking culture of the Bay Area is a self-fulfilling prophecy, drawing in free-thinkers and repelling conservative thinkers in a never-ending virtuous cycle.

> *"Align your interests and passion with the opportunity."*

Charlotte Yates

Angel Investor, Venture Capital Partner, Entrepreneur

Seraph Group, Sumeru Equity Partners, Yates Ltd, Contract Strategies Inc.

What did you want to be as a child?
A doctor.

What was your first career?
Music Production Manager.

What other careers have you had before your current one?
Besides entertainment, commercial real-estate broker, telecommunications executive, IT executive, serial entrepreneur.

Why do you choose to focus on early stage technology investing?
I enjoy working with entrepreneurs, expanding the applicability of their ideas and connecting dots that enable traction for transformative technologies inside of enterprises.

When your investing is not going well, what do you do or say to yourself to keep motivated?
I remind myself that each investment is an opportunity to learn and expand my relationships and experience, which are core elements that enrich my life.

What do you love most about what you do now?
I love bringing together creative, pragmatic solutions to seemingly complex problems and surrounding myself with inspiring people, who are passionate about moving the needle.

What dreams do you have for the next 10 years?
My dream is to inspire and create partnerships that truly transform the future. I will accomplish this through business endeavors, social avenues and personal adventures. I would like to do this in combination with my dream of travelling six months per year to new locations and engaging in projects and activities that support innovation and environmental conservation.

If you could change one thing about the world we live in, what would it be?
I would stop the destruction of the environment, preserving as many animal species as possible, while working on developing sustainable solutions for the future

Can you share a story of one of your worst investments and why it went so badly?
My worst investment was in a fiber-optic technology and services company. The founder had successfully launched several companies in his past, but the companies were in adjacent applications of fiber. The idea resonated with me and I knew that he had executed successfully in his past, so I did not go deeply into the particulars of his roadmap for execution. Over time, I realized that he had not taken a number of risk factors into consideration and was reticent to refine his strategy, based off of pertinent industry data and feedback. The delay in his decisions was devastating to the business. He was afraid to fail fast and pivot. The most amazing accomplishment was that it only took 7 months for them to lose all of their investors' money, including mine.

There is a lot more focus now by investors, employees and customers on companies that do good (and do no harm) as well as make money. How do you think about this topic in the context of your investing?
In order to be comfortable investing in "do good" or socially responsible funds, I think you really need to "want" to be in them for personal reasons. They are more nuanced, generally more expensive than other funds and require more personal focus. While I have not invested in this area previously, I am in the process of evaluating opportunities. My first lesson has been that many financial advisors are nuanced on the differences between ESG, SRI and impact investments, necessitating more personal research. Additionally, in fund selection, I will need to invest

more time in clearly understanding individual investment strategies, as there are subtleties that are not always obvious between funds. There are also additional fees/ expenses associated with this category for multiple reasons. So, why would I invest? Because these types of ventures speak to my values and passion to support innovation and ecological conservation.

Who have been role models you most admire?

My two role models were my mother and concert promoter, Bill Graham. Mom was a single parent with disabilities that persistently pursued her dreams never stopped learning, while always contributing to a higher purpose. Bill taught me the art of ego navigation, creative problem resolution and always delivering outstanding outcomes.

What do you believe are the most important traits of a technology entrepreneur?

Have strong vision and passion, perseverance and fearless about pushing conventional boundaries.

If you could tell a technology entrepreneur just one thing, what would it be?

Ideas are interesting. Clear business models with a route to profitability drive success. Clearly understand your business proposition and target audience and thoughtfully apply and update risks to continuously refine the model.

What do you know now that you wish you had known earlier on?

When I started my first business, I did not apply any creativity to funding. I was fearful of losing control when considering the VC route and therefore decided to grow it slowly and organically. There are so many other ways to fund a business that I never considered, such as innovation funds, accelerators, grants, etc. I could have accelerated much faster with outside funding that would have still allowed me to apply a thoughtful, planned approach to growth and exit.

Would you please create a personal quote that captures important wisdom about participating in early stage investing?

"Clearly understanding your motivations for early stage investing can help in navigating your investment strategy and avoiding disappointments. Investor motives can cover a wide variety of interests such as fulfilling a desire for mentoring, continuous learning, board involvement or simply making money. Align your interests and passion with the opportunity."

Let your curiosity guide you.

> *"Do NOT let your successes or failures define you. They are the outcome of your work, they are not YOU."*

Magdalena Yesil

Founder, Venture Capital Partner, Angel Investor

Zuora, Smartsheet, SoFi, Informed, Broadway Angels, Operator Collective

What did you want to be as a child?
Carpenter.

What was your first career?
Semiconductor Design Engineer.

What other careers have you had before your current one?
My career mirrors Silicon Valley.

Why do you choose to focus on early stage technology investing?
Because I like to live vicariously through other entrepreneurs and I cannot start every company myself.

When your investing is not going well, what do you do or say to yourself to keep motivated?
The only thing I will lose is money.

What do you love most about what you do now?
Working with young people.

What dreams do you have for the next 10 years?
Riding the wave of the next innovation in technology, being more international in my outreach, expanding beyond information technology.

If you could change one thing about the world we live in, what would it be?
Make men get pregnant and give birth.

What was your first technology investment and what happened?
My first investment ever – Securify. Securify was an enterprise software company focused on information security. Invested at a $7M valuation, sold Securify two years later at a $100M valuation. Two founders, both first time entrepreneurs. After three years at the acquiring company, Securify founders spun themselves out, raised venture money and did it again.

Can you share a story of one of your best investments, and why it went so well?
Salesforce. I am the first investor and founding board member. It went well because:

1. We had a successful competitor we were out to kill, so we could be laser focused and targeted.
2. The dotcom bust and recession helped, IT no longer had big budgets and had to look at less expensive alternatives to enterprise software, like SaaS and Cloud offerings.
3. Simple easy to use product that could be offered as a self-service to a well-targeted user base in the enterprise
4. Putting the customer first, glorifying the customer and using the customer as our most valuable sales tool.
5. Very strong sales and marketing DNA

Can you share a story of one of your worst investments and why it went so badly?
IEscrow. In the early days of Ebay, before the seller rating system was fully established, IEscrow provided an escrow service for the buyer and seller to escrow payment until the goods were delivered. Ebay was expected to take it to sellers and buyers, promoting it heavily. They did not. They ended up purchasing the company for pennies on the dollar. Moral of the story – relying on one major partner for your whole revenue stream makes you very vulnerable.

In your opinion, what is so special about Silicon Valley?
Silicon Valley is the land of meritocracy. Does not matter what you look like. Silicon Valley values performance, giving the best and the brightest a chance. It is the land of immigrants - people arrive to make something of them selves. It does not punish failure, instead looks at it as a learning experience. It is iterative in its basic culture, innovation does not happen overnight. Financial gains are a sideshow, not the main focus.

Why do you think the innovation economy/Silicon Valley has a poor track record on diversity and inclusion? What needs to change?

I do not think this statement is true. Look at the people walking on the streets of Palo Alto or San Francisco.

There is a lot more focus now by investors, employees and customers on companies that do good (and do no harm) as well as make money. How do you think about this topic in the context of your investing?

Doing No Harm is paramount and we need to make it job #1. Destroying the environment, working with vendors who use child labor, who do not respect human rights or human health have to be avoided at all cost.

Doing Good is more complicated. "Doing good" is nice to have, and a very important culture to instill in the employees' spirit. At Salesforce we started the 1-1-1 giving program through Salesforce.org. Zuora also has a giving arm. But that does not take away from the need to bring shareholders value by the company's financial performance.

Who have been role models you most admire?

Dan Lynch, Irwin Federman, Sandy Robertson. All three built their relationships strictly on Trust.

What is the best advice you ever received about how to succeed in Silicon Valley?

Let your curiosity guide you.

What do you believe are the most important traits of a technology entrepreneur?

The best entrepreneurs I have worked with have been product people, who have an innate ability to see what a customer would benefit from, create clear specs, build the product and take it to a well-defined market. Never tiring, obsessed with making their dream come true.

What do you believe are the most important traits of an early stage technology investor?

Independent thinker with an ability to take risks.

If you could tell a technology entrepreneur just one thing, what would it be?

If you want mental stability, go get a job. If you want to be an entrepreneur and do your own thing, make sure you are prepared for the extreme emotional ups and down. Surround yourself with friends who can keep you balanced as you swing from thinking you are the smartest person in the world to thinking you are totally worthless and a loser.

What do you know now that you wish you had known earlier on?

Life is short - do not sweat the small stuff.

Would you please create a personal quote that captures important wisdom about participating in early stage investing?

"Do NOT let your successes or failures define you. They are the outcome of your work, they are not YOU."

I wish there was more focus, efforts and dollars dedicated to making sure our planet remains inhabitable and a safe place for generations to come.

"A great company may not be a great investment, especially if you invest too late. A great new technology may not be a great investment if it never gains a market."

Keith Zachow

Angel Investor and Service Provider

Keiretsu Forum, Omega Valuations, CPA

What did you want to be as a child?

I wanted to be my own version of Bo Jackson and play professional baseball and football.

What was your first career?

After undergrad my first job was as a personal lines insurance underwriter, then promoted to an analytics manager because I knew how to use these new things called 'personal computers' then to Senior Financial Analyst. All of which got me into Wharton for grad school.

What other careers have you had before your current one?

Head of investment banking and research at a Canadian broker dealer, acquired and ran a log and burl wholesaler in Oregon, ran a small private equity firm.

Why do you choose to focus on early stage technology investing?

I still find it far more interesting, dynamic and fun than investing in other asset classes.

When your investing is not going well, what do you do or say to yourself to keep motivated?

I ask myself "What can I learn from this?" I find it better than beating myself up over bad returns.

What do you love most about what you do now?

I have wonderful clients and am involved with companies in a broad cross section of the economy, which keeps things constantly new and fresh.

What dreams do you have for the next 10 years?

I've created an ice cream product (CoolBeans®) which I've been trying to launch but Covid had other ideas, so I'd like to get that launched or sold over the mid-term. If fate agrees, I should have some liquidity events from various investments and properties over the next decade too. Beyond that I pray for good health, good friends and meaningful relationships.

If you could change one thing about the world we live in, what would it be?

I'd like to see more geo-engineering research related to climate, such as iron seeding of the oceans, in concert with a global focus on cleaning up the world's oceans and protecting fisheries.

What was your first technology investment and what happened?

I invested in several early-stage technology companies in the early 1990s while doing investment banking in Canada. I soon learned that the quality of management wasn't to my expectations, and they all struggled. It drove me down the coast to Silicon Valley where there were more great jockeys (capable entrepreneurs with great track records) and more great horses (early stage tech companies) than I could keep track of. My early failures in investing simply drove me further to find new opportunities in which to take another shot.

Can you share a story of one of your best investments, and why it went so well?

It wasn't really an 'investment.' I'd moved to the Bay Area in 1995 after quitting my job in Canada. I didn't know anyone here, and my only social life revolved around the Wharton alumni club. One day I saw an email from a fellow grad saying her brother was a scientist at Bell Labs and needed help putting together a business plan for a new invention he created, a wave division multiplexer for use in fiber optics. Of the roughly 2,000 local Wharton grads that received the email, he received one reply—mine. I spent maybe 20 to 30 hours with the founder, helping him with financial projections and investment presentations, in exchange for a percentage of the equity. To make a long story short, the company was acquired by Intel in the early 2000s. With this investment I learned the importance of getting involved early.

Can you share a story of one of your worst investments and why it went so badly?

Well I have plenty to choose from. My worst investments have been companies I hurriedly invested in, whether due to excitement about the technology or FOMO, but didn't pause long enough to have a clear understanding of an exit or what would drive my returns. I'll still take flyers on new investments, but limit the amounts and my expectations.

In your opinion, what is so special about Silicon Valley?

There is no singular thing in my opinion. It is an amalgam of all the components needed to launch great companies: plenty of money and smart investors, incredibly smart entrepreneurs driven to succeed, continuing waves of new technologies, and great weather make it the most exciting place I've ever lived.

Why do you think the innovation economy/Silicon Valley has a poor track record on diversity and inclusion? What needs to change?

With all respect, I see Silicon Valley as a meritocracy—a stark, brutal meritocracy where a few great ideas are rewarded and so many not so great ideas disappear into the abyss. While one may argue that there is limited diversity that there could be more inclusion, I simply don't see exclusion of great ideas, certainly not if there is money to be made. I see it as a real disservice to anyone hoping to succeed in Silicon Valley to go in thinking that they are at a disadvantage because of race or gender. That simply sets them up to fail. It can be rough to start a company. I've seen many wealthy Ivy educated white males reduced to tears when their ideas or their companies failed. Perhaps there is still some discriminatory overhang, but my experience is that when you are talking about startups/funding/investing/exiting, no one really cares about your race or gender.

There is a lot more focus now by investors, employees and customers on companies that do good (and do no harm) as well as make money. How do you think about this topic in the context of your investing?

Well I certainly wouldn't want to invest in a company that did harm to people or the planet, but investing can be very subjective, and what some people call 'harm' others will call it 'good'. My focus is on returns, and when social investing generates greater returns, I'll be more interested in it. Being successful and 'doing good' are not mutually exclusive but neither are they dependent upon each other.

Who have been role models you most admire?

My entrepreneurship professor at Wharton - Ed Moldt - and as corny as it sounds, the founding fathers of this great country.

What is the best advice you ever received about how to succeed in Silicon Valley?

There is always going to be another deal or another investment coming in the future. No one deal is the be-all and end-all. Give your best efforts to whatever deal you're involved in. There are more coming.

What do you believe are the most important traits of a technology entrepreneur?

Deep knowledge of the technology and the market. You can hire good people in finance and marketing and production, but they only succeed if there is someone with complete knowledge of the company's technology and how it fits in the current and coming markets.

What do you believe are the most important traits of an early stage technology investor?

Curiosity and patience and the ability to shake off and ignore failed investments.

If you could tell a technology entrepreneur just one thing, what would it be?

In order for you to make money, your investors also have to make money. It's a competitive marketplace for ideas and funding. Understand your own and your investor's IRRs, and know that your ownership will be diluted, but your overall wealth should grow as you have a smaller % of a much larger pie.

What do you know now that you wish you had known earlier on?

There are no failures, only learning.

Would you please create a personal quote that captures important wisdom about participating in early stage investing?

"Your job is to make investments, and it is important to have a singular focus on that, not focus on great companies or great technology. A great company may not be a great investment, especially if you invest too late. A great new technology may not be a great investment if it never gains a market."

"Practice curiosity."

James Zhang

Entrepreneur, Angel investor, and Advisor

Concept Art House, Fifth Era, Blockchain Coinvestors, Keiretsu Capital

What did you want to be as a child?
I wanted to be an artist as far back as I can remember. I just loved drawing dragons and robots.

What was your first career?
I was a concept artist at George Lucas' video game company: LucasArts from 1998 until 2003.

What other careers have you had before your current one?
I was a concept artist and art director for the video game industry for years before founding my company as CEO/entrepreneur. I got into tech advisory and investments later in my career.

Why do you choose to focus on early stage technology investing?
I love the early stage impact I can make when the entrepreneur's story and ideas line up with my own investment thesis.

When your investing is not going well, what do you do or say to yourself to keep motivated?
"Do the NEXT right thing." I determine what I have control over and what I don't. Then I act accordingly, one problem at a time.

What do you love most about what you do now?
I love playing a small part in a good origin story, whether it's offering advice, connections, or capital to a deserving dreamer or builder.

What dreams do you have for the next 10 years?
Drive change. Create art. Enjoy life. In my 20s, I loved my career as an artist, helping to create fantasy and virtual worlds. In my 30s, I focused on my career and professional growth as a CEO, investor, and advisor. Now, in my 40s, I still love the tech and games industries I work in, but service and impact work has become very important to me. In the next 10 years, I'd like to better merge my 3 passions: Art, Technology, and Impact. I imagine this to be some form of technology-enabled storytelling, but time will tell.

If you could change one thing about the world we live in, what would it be?
At the time of this writing, there is tremendous divisiveness in America. Racial tensions, joblessness, gender and social inequalities feel like a boiling point. There is a worldwide pandemic and the US response has been grossly inadequate. It's hard to point to one change that can improve on all these challenges. I would like to see a change in the US leadership and political structure. I'd like to see leadership that embraces empathy and unity. Leaders who encourage curiosity, equality, and diversity. An administration that listens to science and protects its citizens from deadly outbreaks. Healing has to start.

What was your first technology investment and what happened?
My first tech investment was my own company. I put my own meager capital and sweat equity into my video game and graphics company in 2007. We bootstrapped as a service provider with a big dreams for creating original IP. After attracting some big brands in gaming to become our clients, we brought in our first angel investor in 2008. The next 4 years saw exponential revenue growth and our headcount went from 5 people in a tiny one-bedroom apartment in Shanghai to a team of 130+ in across US and China in 2012. There was a tremendous amount of hard work and challenges in growing a service-based entertainment company. At one point, we were courted by one of Silicon Valley's most successful VCs for a multi-million dollar investment. After a year of courtship, the VC passed on us and I was devastated. Our chairman of the board (The author of this book), took me out to dinner and told me it was okay. There was more than one way to win in the industry. We got back to work. Luck, hard work, networking, and providing great service for our clients led us to a successful acquisition to a public company in the UK. My investors were happy with their returns on their investments. The company is now private once again and doing well. The friendships and capital I made in the sale changed my life and opened the door to me investing and advising in other companies.

Can you share a story of one of your best investments, and why it went so well?

In 2015, I was asked by a long-time friend and client of mine to advise his new mobile game company. He and his partner were a 1-2 punch of game developer and a genius engineer. From years of knowing and working alongside each other, I already knew we'd get along personally. The founders both had sold their previous company. At that company, they made a breakout hit game (500M+ life time revenues). Their thesis for their new game and new company was to bring new innovations in graphics, engagement, and design into an already mature puzzle game genre, of which they were already experts in.

Let's review the ingredients:

1. Founders already had an exit

2. Founders were already successful developers

3. Founders had plan to do something they've done before, but with 20% more real juice.

4. Founders were AWESOME people. Fun and high integrity.

Becoming a shareholder was the easiest of professional decisions for me to make. After a 4+ year grind of ups and downs, switching publishers, a few unsuccessful games, the founders found their way. They built a game that was fun, engaging, and hugely profitable. In 2020, during Covid-19 pandemic, this company was acquired by one of the biggest mobile game giants in our industry. It was a good outcome for all parties. When presented with those ingredients of a proven team, prior exit experience, and a great idea, that's a winning investment recipe.

Can you share a story of one of your worst investments and why it went so badly?

In 2017, I invested in a virtual reality streaming platform company. The founders were 2 brilliant engineers and had an impressive product demo. Their idea was like a Youtube for VR. Ambitious idea for 2017, when Oculus and HTC Vive sales were on the rise. I did months of diligence on the founders. They were passionate and smart, but first-time entrepreneurs. They assembled a team where I knew some of the staff very well, both their strengths and shortcomings.

Here're the ingredients:

1. Brilliant engineers as founders

2. First time entrepreneurs

3. New field with traction and volatility – VR

4. Mixed review on the team.

5. Impressive product demo

I brought in some fellow investors to take a closer look at the deal. We decided to invest as a group, ready to overlook some business concerns in favor of the technology opportunity. We saw signs of trouble less than a year later. The product was behind. The VR market didn't take off as we'd hoped. There just wasn't enough good content and user base in VR. On the management side, some of the investors and I felt the leadership team thought too much alike rather than appropriately challenging one another. The hard meetings began to feel like entrepreneurs vs. investors rather than us all problem-solving together.

Ultimately, the company wasn't able to survive the market challenges and management decisions. The company closed and we lost our money. But valuable lessons were learned about not over-valuing technology above team and market.

In your opinion, what is so special about Silicon Valley?

Of the 5 most valuable companies in the world, we have 3 of them here in Silicon Valley within 20 miles of each other – Apple, Google, and Facebook. I can't stress enough the amount of talent, skill, and knowledge that is discovered, cultivated, and commoditized here in the San Francisco Bay area. This isn't by coincidence. These companies didn't appear over-night. In addition to a disruptive idea, assembling a brilliant team, these companies were supported by some of the most daring and forward-thinking investors in the modern age. Many of these investors like Sequoia, Andressen Horowitz, and Benchmark are in Silicon Valley. For me personally, I'm a video game guy. There are more than 300 game developers and publishers in the SF Bay area. EA, Twitch, Unity, and Zynga have headquartered here. Why are they here? Because the access to talent and capital here is unique and world-class.

This DNA for innovation and venture capital is what makes Silicon Valley the central hub for disruptive technologies such as AirBnB, Salesforce, and Uber. I've helped raised capital for companies in Europe, Asia and US. There's no other place like Silicon Valley in the world.

Who have been role models you most admire?

George Lucas. He gave us Star Wars! (And me, my first job). He pioneered the movie FX industry. He negotiated with FOX to give up director's fees on

Star Wars in exchange for ownership of licensing, merchandising, and sequels remain one of the most profitable business decisions in entertainment history.

What is the best advice you ever received about how to succeed in Silicon Valley?

Kevin Chou, former CEO of Kabam pointed me to an article written by the legendary Ben Horowitz called: *"What's the most difficult CEO skill? Managing your own psychology."*

In summary, it highlights that the 2 biggest mistakes CEOs make:

1. They take things too seriously
2. They don't take things seriously enough.

I know! What are you supposed to do? Short answer: SUCCESSFUL CEOS STAY WITH IT AND DON'T PUNK OUT. I highly encourage anyone interested in being an entrepreneur or investing to read this article and Ben's writings.

What do you believe are the most important traits of a technology entrepreneur?

My most successful entrepreneurial friends have these traits in common:

1. They ask insightful and difficult questions.
2. They are obsessively focused and relatively free of distractions.
3. They have BIG ideas and can articulate WHY their ideas will work.
4. They balance humility with confidence. Our conversations tend to focus on their shortcomings and past failures rather than success stories.
5. They live healthy lives, balancing good diet, exercise, and good mental health, despite the demands of the job.

What do you believe are the most important traits of an early stage technology investor?

My most successful investor friends have these traits in common:

1. They ask insightful and difficult questions.
2. They invest in areas where they themselves have experience in.
3. They have an incredible network of thought leaders and experts.
4. They are genuinely helpful. They know relationships matter in the investment game.
5. They understand "appropriate risk". A friend of

mine does corporate development for a billion dollar gaming business told me: *"My job is basically to say 'No'".* Is he risk averse? Of course! But investment is all about risk. Good investors approach investing with the right amount of risk appetite.

If you could tell a technology entrepreneur just one thing, what would it be?

"Work with people smarter and better than you."

I certainly didn't invent his phrase. The smartest and most accomplished people I've met have said this to me. Jason Citron, founder of Discord was on my advisory board. I learned from him that he wanted to connect the world through mobile devices. His game companies became clients of mine. Jason's quest for disruption was a huge inspiration for me in how I looked at investing. I found myself asking, *"Who is the next Jason Citron? What would Jason do?"* Knowing and working with Jason made me a better entrepreneur and investor.

What do you know now that you wish you had known earlier on?

"Choose powerfully."

Ambition is a powerful thing. We strive for greatness and often fall short of our expectations. We begin each company or investment with dreams of driving change or making lots of money. What happens if we fall short? Do we second-guess our decisions or our self-worth? How many times did I burn out? Wish I stayed an artist? I felt tremendous anxiety and pain so many times in my career before I learned to embrace my journey. Instead of FOMOing (envy) others who are more accomplished, I learned to practice gratitude and choose powerfully my own path.

Would you please create a personal quote that captures important wisdom about participating in early stage investing?

"Practice curiosity."

Team, perseverance, killer ideas are all important. But wisdom starts with curiosity. We "Practice", because there's no mastering curiosity. There's only asking better questions. Have a great idea? Ask around. See if it resonates. Want to impress in a pitch meeting? Ask amazing questions. Most importantly, practice curiosity with yourself. "Why is this idea important? How do we win? Who will build this? Who else is doing this?"

Contributor
Index

Acknowledgments

We should start the acknowledgments by recognizing that it was Grace Bonney's fabulous book *In The Company of Women* that sparked the flame that became this project. Grace, imitation is the sincerest form of flattery. Thank you.

Secondly, we recognize the brilliance of Benjamin Graham who wrote the original *The Intelligent Investor* more than 70 years ago. We pored over every page of your book while at Stanford University Graduate School of Business and in the years that followed, and we appreciate your very hard work. We just wish that in 1949 there were more disruptive technology companies to include in your analysis.

A special thank you goes to all the people who encouraged us to move forward with this project and of course the 50 or more who found the time to put your wisdom down on paper and allow us to include it in this book. You are among the elite investors and advisors of Silicon Valley and we know just how precious your insights and advice are.

As to the writing of the book, Alison Davis and Matthew C. Le Merle authored this book, and all errors and omissions are theirs alone. Fifth Era Media was our publisher. Andy Meaden designed the book's interior layout and the covers and Leonardo Q. Le Merle compiled it.

Thank you for buying it and please send you feedback to us – good and bad, we will surely learn from it.

Alison Davis
Matthew C. Le Merle
San Francisco, California, USA

About the Authors

Alison Davis

Alison Davis is co-founder of Fifth Era. She is an experienced corporate executive, public company board director, an active investor in growth companies and a bestselling author on the topics of technology and innovation. Currently she serves on the boards of Collibra, Fiserv, and Silicon Valley Bank, and chairs the advisory board of  Blockchain Capital. She was CFO at BGI (Blackrock), managing partner at Belvedere Capital, and a strategy consultant at McKinsey and A.T. Kearney. Alison has degrees from Cambridge (MA/BA) and Stanford (MBA). She was born in Sheffield, England, and has lived for the last 25 years in the San Francisco Bay Area where she raised her family with her husband, Matthew C. Le Merle. For more information, go to **www. alisondavis.com**.

Matthew C. Le Merle

Matthew Le Merle is co-founder and Managing Partner of Fifth Era and of Keiretsu Capital—the most active early stage venture investors in the world.

Matthew grew up in England before living most of his life in Silicon Valley where he raised his five children with his wife, Alison Davis. Today he splits his time between the United States and the UK. By day he is an investor in technology companies, manages Blockchain Coinvestors, and is a bestselling author and speaker on innovation, investing and the future. In his spare time, he enjoys reading, writing and photography. He was educated at Christ Church, Oxford and Stanford University and is an adjunct professor at Singularity U. For more information, go to **www.matthewlemerle.com**.

Disclaimers

Income Disclaimer

This document contains recommendations for business strategies and other business advice that, regardless of our own results and experience, may not produce the same results (or any results) for you. We make absolutely no guarantee, expressed or implied, that by following the advice in this book you will make any money or improve current profits or returns, as there are many factors and variables that come into play regarding any given business or investment strategy. Primarily, results will depend on the nature of your due diligence, product, or business model, the conditions of the marketplace, and situations and elements that are beyond your control. As with any business endeavor, you assume all risk related to investment and money based on your own discretion and at your own potential expense.

Liability Disclaimer

By reading this document, you assume all risks associated with using the advice given herein, with a full understanding that you, solely, are responsible for anything that may occur as a result of putting this information into action in any way, regardless of your interpretation of the advice.

You further agree that neither we, nor our companies can be held responsible in any way for the success or failure of your business or investments as a result of the information presented in this book. It is your responsibility to conduct your own due diligence regarding the safe and successful operation of your business or investment portfolio if you intend to apply any of our information in any way to your business or investment operations.

Terms of Use

You are given a non-transferable "personal use" license to this product. You cannot distribute it or share it with other individuals without the express written permission of the authors. Also, there are no resale rights or private label rights granted when purchasing this book. In other words, it's for your own personal use only.

Affiliate Relationships Disclosure

We make a number of references in this book to entrepreneurs, companies, or programs that we have invested in, worked with, or recommend. We have no paid affiliate relationship at all with any entrepreneur, company, or program we reference with respect to inclusion in this book.

Visit

www.Tiisv.com

to receive additional content about topics in this book,
and to find out more about the authors.

CPSIA information can be obtained at www.ICGtesting.com
Printed in the USA
LVIW011914121120
670524LV00002BA/2